Fundamental Theories of Physics

The international monograph series "Fundamental Theories of Physics" started with a new Editorial Board from volume number 172 on. Earlier volumes which appeared in the names of the new Editorial Board were still reviewed and published under the responsibility of the previous Editors of the series.

Volume 172

For further volumes:
http://www.springer.com/series/6001

The series "Fundamental Theories of Physics" aims at stretching the boundaries of mainstream physics by clarifying and developing the theoretical and conceptual framework of physics and by applying it to a wide range of interdisciplinary scientific fields. Original contributions in well-established fields such as Quantum Physics, Relativity Theory, Cosmology, Quantum Field Theory, Statistical Mechanics and Nonlinear Dynamics are welcome. The series also gives a forum to non-conventional approaches to these fields. Publications should provide perspectives and carefully bridge conventional views with the presented new and promising ideas.

Although the aim of this series is to go beyond mainstream physics, a high profile and open-minded Editorial Board will carefully select the contributions and will ensure the high scientific standard of this series.

Laura Mersini-Houghton • Rudy Vaas
Editors

The Arrows of Time

A Debate in Cosmology

Springer

Editors
Prof. Laura Mersini-Houghton
University of North Carolina
Dept. Physics and Astronomy
Chapel Hill North Carolina
USA
mersini@physics.unc.edu

Rüdiger Vaas
bild der wissenschaft
Ernst-Mey-Str. 8
D-70771 Leinfelden-Echt.
Germany
Ruediger.Vaas@t-online.de

ISBN 978-3-642-23258-9 e-ISBN 978-3-642-23259-6
DOI 10.1007/978-3-642-23259-6
Springer Heidelberg Dordrecht London New York

Library of Congress Control Number: 2012939849

Printed on acid-free paper

Springer is part of Springer Science+Business Media (www.springer.com)

Contents

Introduction

Time is of central importance to science and philosophy. And yet, the simplest questions – "is time real, or is it an essential part of the structure of human intellect?" – remain largely controversial. Theories of nature can be broadly categorized into two sets of information: physical laws which relate the sequence of states of a system, and initial conditions determined at a fixed moment in time. Clearly, a description of the succession of states or the choice of an initial moment where data about the system is defined, also involve time. Understanding this turns out to be as difficult as probing the origins of the universe since both physical laws and initial conditions assume a concept of time, which, in most cases, is inseparably interwoven into the theory and its predicted outcomes. Though time is ubiquitous and intuitive, it still defies comprehension. Disentangling ourselves from time to enable objective and independent investigation is the challenge.

Understanding the nature and the direction of time has occupied the minds of philosophers and scientists throughout history. It continues to do so. As far back as the fifth century, St Augustine wrote in *Confessions*, Book 11: "what then is time? If no one asks me I know. If I wish to explain it to one who asks, I know not... My soul yearns to know this most entangled enigma". The enigma persists. The yearning for understanding has now fallen on physicists working on the most fundamental questions about the cosmos and the origins of the universe.

What is the enigma of time?

1. The nature of time: is time an inherent and intrinsic ingredient of nature or did it emerge only at the big bang?
2. The arrow of time: why is there a clear direction from past to future, i.e. what breaks the time translation symmetry and sets an arrow of time? Why should the birth of the universe be determined by this direction?
3. The time-symmetry of physical laws: why is it that the laws of physics, which describe the universe we live in, cannot distinguish between past and future? How can physical laws respect time translation symmetry when the universe breaks it at the big bang?

L. Mersini-Houghton and R. Vaas (eds.), *The Arrows of Time*, Fundamental Theories of Physics 172, DOI 10.1007/978-3-642-23259-6_1,

This book describes contemporary views of physicists on the nature of time, the origin of time translation symmetry breaking, and the implications this enigma has on the origin and predictions of physical laws.

A consistent treatment of the three time enigmas described above, is presented in Laura Mersini's chapter in the context of the multiverse. The problem is addressed by taking the view that the "local" time in our universe should be distinguished from the fundamental time of the multiverse. The multiverse is a closed system, thus it preserves the time symmetry. Our universe is an open subsystem and the event of its birth breaks the time symmetry locally creating an arrow of time in its domain. Since the physical laws in our universe are inherited from the multiverse then it follows that they are time symmetric despite the fact that the whole universe has broken the time symmetry from the moment of the big bang.

An overview of the many arrows of time and their connections is given by Rüdiger Vaas. He discusses different explanations for their origin and focuses on useful conceptual distinctions. Furthermore he suggests a multiverse framework in which the (or our) big bang created the arrows of time. In this framework, the big bang might have originated as some sort of pseudo-beginning in a quantum vacuum that has no direction of time (macrotime) but nevertheless some sort of symmetric microtime. It is even possible that time ends – although paradoxically, it may do so only temporarily. Some recent cosmological models do in fact instantiate such pseudo-beginning and -ending scenarios.

The close correlation between timekeeping devices and physical laws is addressed both in the chapter by Rodolfo Gambini, Rafael A. Porto, and Jorge Pullin and in that by Andreas Albrecht and Alberto Iglesias. Albrecht and Iglesias focus on the general case of the ambiguity associated with the choice of clocks, leading to the immediate implication that we cannot have a fixed set of laws since different clocks would lead to different predictions of the theory. The view taken there is that the problem of clock ambiguity may be bypassed if physical laws emerge statistically from a random time-independent Hamiltonian.

Gambini, Porto, and Pullin present crucial aspects of the impact clocks have on quantum theory, specifically the measurement problem. Devices that measure time, like all quantum systems, are subject to quantum fluctuations. Therefore they are constrained by a fundamental bound on the precision of their timekeeping ability. The intrinsic uncertainty of clocks, given by this bound, makes it unrealistic to expect an accurate, deterministic measurement of time in quantum systems, including physics near the big bang or a black hole.

Martin Bojowald tackles the issue of clock ambiguity and its quantum uncertainty, based on a phenomenological model inspired by quantum gravity, whereby changing clocks is equivalent to a gauge transformation.

Several implications of time's arrow, along with proposals that circumvent this problem, are presented in the chapters by Tom Banks, by Gary W. Gibbons, by Katherine Freese, Matthew G. Brown, and William H. Kinney, and by H. Dieter Zeh. The origin of the cosmic arrow of time is closely related to the origins of the universe by the second law of thermodynamics. According to this law, the universe must have started in an incredibly ordered (low entropy) state in order to be consistent with the observed time's arrow.

The possibility that the origin of the arrow of time is rooted in quantum cosmology is presented in the chapter by Claus Kiefer. This origin would be a consequence of imposing low entropy boundary conditions on the wavefunction of the universe. A low entropy boundary condition is a natural choice since the decoherence process increases the entropy of the wavefunction in an irreversible manner.

Tom Banks advocates applying the holographic principle to the Boltzmann–Penrose question of why our universe started in such a low entropy state, while Freese, Brown, and Kinney provide a concrete model of "phantom bounces and oscillating cosmology" in which the universe is naturally driven through low entropy states at the start of each cycle. Gibbons proposes a sharp form of Thorne's hoop conjecture for the formation of black holes, which relates Birkhoff's invariant to the ADM mass of the outmost apparent horizon.

An interesting question related to time's enigma is the puzzle of why different arrows of time, such as for example, the cosmic arrow determined by the expansion of the universe, the biological arrow determined by (say) human ageing or the thermodynamic arrow determined by the increase in entropy, agree with each other. H. Dieter Zeh makes the case that the "Master Arrow of Time" (the combination of all time's arrows) does not have to be the same as a formal time parameter needed to measure the succession of global states. Zeh also discusses the arrows in both classical and quantum physics, the retardation of various kinds of correlation, the dynamical rôle of quantum indeterminism, and different concepts of timelessness (quantum gravity included).

Reading through the chapters of this book takes us on a fascinating voyage through the diversity of current schools of thought on the very basic question – what is time? A question that remains stubbornly obscure despite centuries of investigation. Time, the entity we are all intuitively wired to acknowledge and take for granted from birth.

As Lord Byron wrote: "Time! The corrector when our judgments err..." Hopefully time will tell which of the judgments presented in this book will stand the test of time.

Time After Time — Big Bang Cosmology and the Arrows of Time

Rüdiger Vaas

Abstract Time, as familiar as it seems to us in everyday life, is one of the greatest puzzles of science and philosophy. In physics and cosmology it is especially mysterious why time appears to be "directed", that is, why there seems to be an essential difference between the past and the future. The most basic known laws of nature do not contain this asymmetry. And yet, several arrows of time can be distinguished – at least ten, in fact. However, it is unclear whether any of them are fundamental or whether others can be reduced to these, and it is not known how the direction of time could be explained convincingly. From the growing but still astonishingly low entropy of the observable universe, it seems plausible that the solution of the mystery is connected with cosmology and an explanation of the big bang. This could require a new fundamental law of nature (which might be related to a particular geometry) or specific boundary conditions (which might be comprehensible within the framework of a multiverse theory). Or it may be that time's direction is fundamental and irreducible, or an illusion and not explicable, but can only be "explained away". It is even more confusing that not all of these alternatives are mutually exclusive. Furthermore, there is a plethora of approaches to explain the big bang. Some models postulate an absolute beginning of time, others an everlasting universe or multiverse in which the big bang is a phase transition, and maybe there are myriads of big bangs. So the low entropy of the observable universe might be a random fluctuation – whereas elsewhere even opposite thermodynamic directions of time may arise. Perhaps the (or our) big bang just created the arrows of time, if it originated as some sort of pseudo-beginning in a quantum vacuum that has no direction of time. Thus it seems useful to conceptually distinguish an undirected microtime and a directed macrotime. It is even possible that time ends – although paradoxically, it may do so only temporarily.

R. Vaas (✉)
bild der wissenschaft, Ernst-Mey-Str. 8, D-70771 Leinfelden-Echt., Germany
e-mail: Ruediger.Vaas@t-online.de

L. Mersini-Houghton and R. Vaas (eds.), *The Arrows of Time*, Fundamental Theories of Physics 172, DOI 10.1007/978-3-642-23259-6_2,

Man is ... related inextricably to all reality, known and unknowable ... plankton, a shimmering phosphorescence on the sea and the spinning planets and an expanding universe, all bound together by the elastic string of time. It is advisable to look from the tide pool to the stars and then back to the tide pool again.

John Steinbeck: The Log from the Sea of Cortez (1951)

Talkin' bout that youthful fountain
Talkin' bout you and me
Talkin' bout eternity
Talkin' bout the big time

Neil Young: Broken Arrow (1996)

1 The Direction of Time

"Time flowing in the middle of the night, / And all things creeping to a day of doom," wrote the British poet Alfred Lord Tennyson. Yet this unceasing stream of time, existing apparently without dependence on its recognition, is perhaps only an illusion – but also a problem. Because the known laws of physics are time-symmetric. So they neither entail nor prefer a direction from past to present.

However, our everyday experience teaches us the opposite. For only processes with a clear direction are observed in the complex systems of nature and culture: blossoms become apples that later decompose; milk drops into black coffee, making it brown; a glass falls from the table and bursts into a thousand pieces. Even cyclic processes of nature such as the seasons or the phases of the moon are parts of irreversible dynamics. Whoever watches mold turning into a red apple, milk drops hopping from a coffee cup, or shards being resurrected into a glass probably would feel like he is in the wrong movie – or simply watching one that is running backwards.

Irreversibility is why – or how – the formation and development of complex structures is much less likely than their decay or something turning into dust and ashes. By use of the concept of entropy this can be quantified physically: it is a measure of a system's degree of disorder. And disorder is much more probable than order. There are, for example, significantly fewer possibilities of molecular combination for a small drop of milk in coffee than for a good mixing. This is why entropy only increases on average, as the second law of thermodynamics states, while the first law expresses the conservation of energy (see [28, 29, 32, 91] for a historical introduction to thermodynamics).

The development of local order does not contradict the second law of thermodynamics. Contrariwise, it creates more disorder within the entire system. In general, entropy does not decrease globally, but can do so locally. Therefore, the formation of complex structures, in other words order, is not impossible, but it occurs only at the expense of a greater amount of disorder in the environment (see, e.g., [65, 66]). Cleaning your desk, for instance, means eating more lettuce, the leaves of which gain their energy from nuclear fusion in the sun – local order increases, yet so does the amount of chaos in the solar system.

So the second law marks a direction of time – or developments in time, which does not necessarily mean the same thing. Yet the second law is not the solution of the problem, but its core. Because all of the known apparently fundamental laws of nature are time-symmetric: they don't include entropy increase; they don't contain a preferred direction of time; they don't differentiate between future and past in principle. This time-reversal invariance means that every macroscopic process could also run in reverse. So why doesn't it in our universe?

This question could be rejected as meaningless if one argues like this: disorder increases with time because we measure time in the direction in which disorder increases. However, this does not solve the problem, because it would still remain unclear why the thermodynamic direction of time exists in the first place. The developments could, after all, also alternate between forward and backward – or not take place at all (see [116]).

It is important to bear in mind that time-reversal invariance and reversibility are not the same but independent from each other and not necessarily correlated (following [3]). Time-reversal invariance is a property of dynamical equations and of the set of their solutions. Reversibility is a property of a single solution of such an equation. Dynamical equations are time-reversal invariant if they are invariant under the application of the time-reversal operator T, which performs the transformation $t \to -t$ and reverses all dynamical variables whose definitions as functions of t are not invariant under this transformation. If $f(t)$ is a solution of such an equation, then $Tf(t)$ is also a solution. These "time-symmetric twins" are temporal mirror images of each other and only conventionally different if no privileged direction of time is presupposed. A solution $f(t)$ is reversible if it does not reach an equilibrium state where the system remains forever. (In classical mechanics, for instance, a solution of a dynamical equation is reversible if it corresponds to a closed curve in phase space.) Time-symmetric laws are, in conclusion, perfectly compatible with asymmetric solutions (see [182]).

2 Ten Arrows of Time

Why do we remember the past, but not the future? For this asymmetry of our experience of time – an irreversibility of many processes and thus the direction of time – Arthur Stanley Eddington [41] coined the metaphorical expression "arrow of time".

It is an open and controversial issue whether there is a single arrow, including and perhaps "guiding" all physical processes, or whether some processes evolve in some sense independently of each other, instantiating their own arrows of time (detailed reviews and elaborations are given, e.g., by [35, 67, 131, 181]).

Conceptually, at least ten different arrows – categories of phenomena that have a direction in time – can be distinguished (see [148]):

- The psychological arrow of time: we remember the past, which seems immutable, but not the future, which isn't fixed for us yet. We experience a "stream" of time that doesn't turn back but moves us from birth to death. The psychological arrow is related to a computational arrow, if cognitive processes are computational – at least partly (omitting issues of phenomenal content aka qualia here).

- The causal arrow of time: effects never precede their causes, and these have coherent structures (at least in classical systems).
- The evolutionary arrow of time: complex natural but also cultural systems are based upon directed developments and often also upon differentiation. Exponential growth can only be observed in self-organizing systems.
- The radioactive arrow of time: exponential growth is confronted with exponential decay of radioactive elements which marks a direction in time as well.
- The radiative arrow of time: electromagnetic radiation diffuses concentrically from a point but never coincides at one point after moving in concentrically from all sides. (This is also true for sound waves, or for waves that result from a stone being thrown into water, or for the assumed gravitational waves emitted by rotating, collapsing, or colliding massive bodies.)
- The thermodynamic arrow of time: the entropy of a closed system maximises, so the system seems to strive for its thermodynamic equilibrium. For example, coffee cools down to ambient temperature and milk drops that have been poured into it don't stay together but disperse evenly.
- The particle physics arrow of time: the decays of certain particles, the neutral K mesons (kaons) and B mesons, and there antiparticles lead implicitly to the conclusion that there is an asymmetry of time because these decays break other symmetries. (More precisely, some processes governed by the weak interaction violate time reversal T, but can also be subsumed under time-reversal invariance nevertheless, because T-violation is compensated by an application of a unitary CP-transformation, and according to the CPT-theorem the combination of charge conjugation C, parity transformation P, and time reversal T is conserved.)
- The quantum arrow of time: measurements – or interactions with the environment (quantum decoherence) in general – interfere with a quantum system which realizes all possible states in superposition, and lead to only one classical state being observed. This so-called collapse of the wave function (if it really happens) describes, for example, why Erwin Schrödinger's infamous cat is not (observed as) dead and alive at the same time. Instead of collapsing, reality could also "split" into different parallel universes that would henceforth be independent of each other, so that all alternatives were simultaneously realized – in one world the cat is dead and in another one it is alive.
- The gravitational arrow of time: gravity forms structures, for example galaxies and stars, from tiny density fluctuations within the almost homogeneously distributed primordial plasma of the early universe (Fig. 1). Gravitational collapse can even create black holes. They are "one-way streets" of matter, places of highest entropy, and perhaps even irreversible annihilators of physical information. This arrow is also called (or subsumed under) the fluctuation arrow [70].
- The cosmological arrow of time: space has been expanding since the big bang.

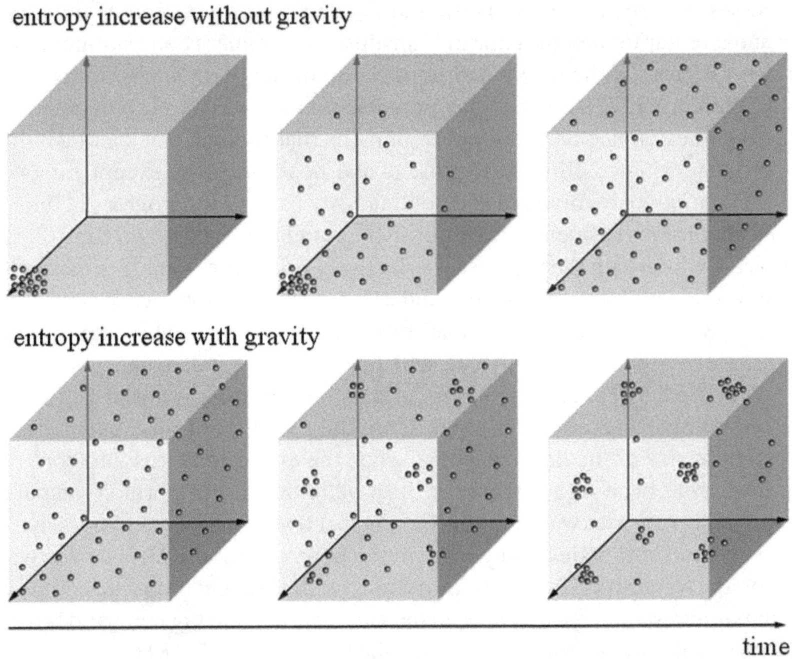

Fig. 1 Growing disorder. Entropy – the physical measure of a system's disorder – can only increase statistically in the course of time. For this reason it even defines in a way the direction ("arrow") of time. If a gas bottle is opened up in empty space, the gas molecules soon spread evenly throughout the entire volume – then, thermodynamic equilibrium is reached as a state of maximum entropy (top). Yet in a large space such as the early universe, gravity creates local concentrations of the originally almost homogeneously distributed gas (bottom) – and this was how stars and galaxies formed. An increase of entropy follows from this gravitational effect, which was not taken into consideration for a long time. There is still a debate about whether there can be a thermodynamic equilibrium, a "heat death", in an expanding space, and how the total entropy in the universe can be usefully defined at all.

These ten temporally directed processes seem more or less unrelated to each other at first glance. Yet given that at least today all arrows point in the same direction, it seems natural to search for a primordial, super, or master arrow of time that all the others could be ascribed to. The particle physics arrow of time, the cosmological arrow of time, and the thermodynamic arrow of time are likely candidates.

The thermodynamic arrow might be responsible for the psychological and the evolutionary arrow of time (cf. [68]; but see [94] for an argument against the correlation of thermodynamic and psychological or computational arrows, respectively). Entropy can also be defined for black holes and hence for gravitational processes (see [85, 113]). Causality is a difficult issue (see, e.g., [128, 129]) taken by some to be subjective or as a logical relation, ultimately, and thus reducible to other arrows (see [117]), but as the "cement of the universe" by others (see, e.g., [92]); thus perhaps causality is not only a pragmatic consideration but grounded in processes governed by conservation laws such as conservation of mass-energy, linear and angular momentum during transmissions of energy and momentum.

Still mysterious is the origin and implication of the particle physics arrow [15]. Joan A. Vaccaro [172] argued that processes which violate T-symmetry induce destructive interference between different paths that the universe can take through time. The interference eliminates most of the possible paths except for two that represent continuously forwards and continuously backwards progress in time. Data from accelerator experiments allow the distinction between the two time directions and indicate which path the universe is effectively following. Thus T-violation might have large-scale physical effects that underlie the unidirectionality of time.

There have also been controversial discussions about whether the arrow(s) of time in an evolving closed universe will reverse in the collapsing phase [35, 54, 55, 67, 71, 72, 82, 107, 112, 113, 181]. Perhaps the different arrows of time reduce to the cosmological arrow, so in some sense the direction of time would switch at the maximum size of the finite universe, when the expansion turns into contraction. Sometimes it has been argued that even the psychological and thermodynamic arrow of time would run backwards (from the perspective of the expanding stage), and observers would still believe they were living in an expanding phase. In a quantum cosmological framework, however, everything with classical properties is destroyed in the maximum stage, due to quantum interference, and the big bang and big crunch are ultimately the same, amusingly called the big brunch [82, 181].

3 Four Kinds of Answers

Where does the asymmetry of time – or at least the processes in time – originate if most laws of nature are time-reversal invariant and thus do not prefer a direction in time? Basically, four kinds of answers can be distinguished [148]:

- Irreducibility. The direction of time is not a derivable phenomenon but an essential attribute of time: then time simply passes and is independent, for example, of entropy. Many philosophers share this opinion. Tim Maudlin [98], for instance, defends it and accuses skeptics of only being able to argue for time symmetry because they already presuppose it. However, this objection might be reversed and Maudlin could be accused of not admitting the problem in the first place.
- Laws. Perhaps there is a fundamental, but still unknown law of nature that is time asymmetric. Accordingly, Roger Penrose [112] hopes that such an arrow of time follows from a theory of quantum gravity that unites quantum theory and the general theory of relativity. This might also explain the mysterious collapse of the wave function that many physicists assume. Therefore quantum theory would have to be modified in such a way that it contains a time asymmetry. Then the past could be calculated from a future perspective but not the other way round. This possibility would help historians to gain an advantage over physicists. Other

researchers, Ilya Prigogine [118] for instance, localize arrows of time in the peculiarities of complex systems far from thermodynamic equilibrium, which are postulated to have special laws.

- Boundary conditions. Most physicists assume that the irreversibility of nature is not based upon time-asymmetric laws but is a result of specific, perhaps very improbable initial or boundary conditions (cf. [4, 131, 181]). The problem would thus be shifted to the origin of the universe and accordingly to models of quantum cosmology, though there is no consensus about the nature and form of these boundary conditions. A subset of such explanations are proposals of cyclicity. Here the universe oscillates through a series of expansions and contractions (e.g. [6, 27, 50, 140]) and/or evolves through a perhaps infinite series of big bangs (e.g. [113], see below). Real cyclic models, which do not shift the problem of time's arrow into the infinite past, have to show how the entropy created in each cycle is destroyed or diluted before or within the subsequent big bang, in order to reset the stage for the next oscillation. Therefore a decrease of entropy or entropy density must be explained.
- Illusion. If time is not objective – a property of the world or at least of some of its objects or their relations – but subjective, physicists are searching for an explanation in the wrong place. Immanuel Kant assumed time to be a pure form of intuition or perception, inherent in the human mind, a kind of transcendental requirement or pre-structure for the possibility of experience itself, hence nothing that belongs to the things in themselves. He claims that "time and space are only sensible forms of our intuition, but not determinations given for themselves or conditions of objects as things in themselves. To this idealism is opposed transcendental realism, which regards space and time as something given in themselves independent of our sensibility" ([80], A 369). Other philosophers suspect time of being a construct of consciousness or of the grammar of our language. There are also powerful, but controversial arguments from physics, especially relativity and quantum gravity, emphazising that there is no time independent from space, only a spacetime unity, or that a time parameter does not even appear in the fundamental equations of quantum gravity (such as loop quantum gravity) or quantum cosmology (especially the Wheeler–DeWitt equation) (see, e.g., [12, 84, 125, 181, 182]). So perhaps the arrows of time do not even exist in the world as such. If the entire history of the universe is there as a whole or unity, time would be a mere illusion in a certain sense.

Some of these accounts are mutually exclusive, others are not. For example time could be an illusion (or emergent), but an asymmetrical block universe (if, say, the big bang has much lower entropy than the big crunch) would still deserve an explanation, which might consist in a specific boundary condition or law. Or if time is fundamental (as, e.g., [31, 102] argue), this might be represented by a law too. Or perhaps specific boundary conditions are really an instantiation of a special law, as suggested by Stephen Hawking [74].

4 Fundamental Issues

It is a deep conceptual, physical, and even metaphysical question whether time is fundamental or not. What does this mean, and how can we know about it?

It is not clear from a conceptual point of view whether the direction of time is a necessary feature of time. If so, and if time is fundamental, then the arrow of time is fundamental too. In this case the chances of finding a deeper understanding, or at least a testable explanation in physics or cosmology, are slim. Time as a fundamental entity, as well as its arrow(s), could of course be represented as a fundamental parameter in a future fundamental theory, but this would be no derivation or explanation, just an assumption. Ultimately, time – and the arrow of time – would remain a mystery. This might very well be the case. However, methodologically, it can and should not be a premise or limit of research. On the contrary, scientists and philosophers alike should try to reduce and/or explain time – and proceed as far as they can get. Even wrong explanations are better than no explanation at all, because they can be revised and improved. And their errors may teach useful lessons nevertheless. If an explanation doesn't work, it could still tell us something new, if it is possible to understand why it doesn't work.

Note also that time could be fundamental whether or not it has a beginning. If time originated with the big bang, there was no "before". On the other hand time might be eternal, thus preceding the big bang, which is compatible with a multiverse scenario producing countless big bangs and, hence, universes. But the view that time is emergent is also consistent with either an absolute beginning or temporal eternity within or without a multiverse.

If time is fundamental, this doesn't imply logically that the arrow of time is also fundamental. Perhaps time's direction requires additional assumptions, such as causality or specific initial conditions, which might not be fundamental and could be explained (or they are purely accidental and therefore not further explainable).

For example, the fundamental theory might include a basic time parameter but still not tell us why the entropy of the universe is as low as it is, nor why time's direction could not change. The fundamental theory might be time-symmetric nevertheless – just as classical mechanics, the theory of electromagnetism, general relativity, quantum mechanics, and quantum field theory are. Alternatively, there could be a fundamental, even eternal time within a multiverse scenario where different universes or parts of the multiverse have different directions of time. Or microtime may be fundamental while macrotime (including an arrow, see [154, 162], and below) is not; thus there may be places without local or even global arrows of time – as there could exist islands of reverse arrows (cf. [131–134], but against this [181]). Perhaps the far future empty universe will approximate to such a timeless place (cf. [170]), or there may already be localized regions somewhere within the universe, or there was such a state before the big bang, e.g., a quantum vacuum. Of course such conceptual possibilities are not solutions of the problem, just surveys, and conclusions must be supported by scientific arguments.

If there is no (or no non-reducible) time parameter in the fundamental theory, which is not known yet, one might argue that time is not fundamental – following the view that metaphysics should be determined or framed by our best scientific theory. This reasoning is controversial. But if one accepts the nature of time as a (at least partly) metaphysical issue at all, then attempts to understand it should be in accord with the best scientific theories. And if the fundamental theory contains no time, as some approaches in quantum gravity already suggest, time might be "emergent" or illusionary indeed. If so, ordinary time – or its many aspects – can (or must) be explained in a certain sense. And then there are good chances that the arrow of time can be explained too – at least approximately.

This explanation need not necessarily be a physical or cosmological one, by the way. Perhaps "time" has something to do with how we are practicing science, that is predict and retrodict events and facts. But this is based upon our everyday thinking, how we deal with our experiences and how we order sensations and intentions; it could have been simply advantageous in the evolution of our cognition and behavior (perhaps even a kind of useful illusion such as believing in free will or deities, see [149, 160]). Thus it might turn out that it is sufficient to take time as an ordinary-life concept, a way to describe and handle sensations and actions, to characterize it phenomenologically, and, perhaps, to search for a neuropsychological (or even neurophysiological) explanation. If so, the riddles of time would not be a genuine part of physics, only inherited by physics or transformed into it, but ultimately solved by cognitive neuroscience (see, e.g., [119, 120, 147]). In this respect, time might even be fundamental to us, together with space, that is "pure forms of sensible intuition, serving as principles of a priori knowledge", as Kant ([80], B 36) put it, and hence experimentally opaque, but for practical reasons transferred as a parameter into scientific theories. Thus time could be both fundamental (for observers) and an illusion (not existing mind-independently) – and even emergent (e.g., arising in complex neural networks of cognitive systems). To avoid conceptual confusion it is therefore important to clarify notions such as "fundamental", "emergent", "reducible", "illusionary", etc. in respect of the scope of application.

Though the concepts of time and its direction are indisputably important for our cognitive setting, it would require strong arguments assuming it is sufficient to reduce questions regarding the arrows of time in physics and cosmology simply to cognitive neuroscience or even philosophical phenomenology. It is trivial that science requires scientists, but it would be a non sequitur to claim because of this that there are no features independently from scientists or conscious states and events in general. It would be very surprising if scientific explanations end in or lead to scientific minds, rather than starting from them.

5 Gravity, Entropy, and Improbability

The second law of thermodynamics results – at least phenomenologically – from there always being more disordered states than ordered states. This can be illustrated by a box with many pieces of a puzzle. There is one and only one arrangement in which the pieces create a picture. Yet there is a high number of combinations in which the pieces are disordered and do not form a picture. This is similar to the molecules of stirred milk in a cup of coffee: theoretically they could agglutinate into a drop; in practice they never do because this is so improbable. The reason for such extremely low likelihoods is not represented by laws of nature, however, but by the boundary conditions, respectively the initial conditions. And it is these that pose a conundrum.

So one can argue like this (see [177]): Why does the thermodynamic arrow of time exist? Because the present entropy is so low! And why is it so low? Because it was even lower at earlier times!

This explanation is, however, as elegant as it is insufficient. Because it only shifts the problem, relocating it to the remote beginning of our universe. Yet the big bang 13.7 billion years ago lies in a dark past – and this is not just meant metaphorically. There was no light until 380,000 years after the big bang, when the universe had cooled down enough to release the cosmic microwave background radiation that we can measure today. Given that this radiation is extremely homogeneous – aside from tiny fluctuations in temperature on the order of a hundred thousandth of a degree – matter must have been extraordinarily uniformly distributed at this early epoch and in thermal equilibrium with the radiation. (Dark matter, if it exists, does not interact electromagnetically, and would have been 10 to 100 times more concentrated.)

The spectrum of the cosmic background radiation today almost perfectly resembles the electromagnetic radiation of an idealized black-body in thermal equilibrium with a temperature of 2.725 K (with an emission peak at 160.2 GHz). This might appear paradoxical at first, given that such an equilibrium is often assumed to be the maximum of entropy – like the heat death of the universe that physicists in the 19th century imagined to be the bleak end of the world, consisting ultimately only of heat and perhaps homogeneously distributed particles, if there are any that cannot decay.

Yet appearances are deceptive: the homogeneous fireball of the early universe did not have a high, but a very low entropy! Because in the balance gravity must not be ignored – something that was not recognized for a long time. And gravity is working in the opposite direction: clumping, not homogenizing. So at large scales homogeneity doesn't show a high entropy, but contrariwise a very low one, because gravity's part of the entire entropy here is very low. The strongest "concentrations" of gravity, black holes, are also the biggest accumulations of entropy. Physically speaking, gravitational collapse leads to the greatest possible amount of disorder. The entropy of a single black hole with the mass of a million suns (such as the one at the galactic centre, for example) is a 100 times higher than the entropy of all ordinary particles in the entire observable universe. Yet the homogeneous cosmic background radiation and further astronomical observations very clearly show that black holes did not dominate the very early universe, and this has remained so until today.

This extreme uniformity of matter distribution and the "flatness" of our universe's spacetime geometry themselves appear almost as a miracle. Penrose [111, 112] was the first to recognize and even quantify this. Compared to all possible configurations of matter and energy in our universe, the actual state is extremely improbable.

Penrose estimated it to be a mere $1{:}10^{10^{123}}$, more recent data imply circa $1{:}10^{10^{122}}$ [85]. This double exponent is unimaginably huge. It has so many zeros that it would, if printed in the format of this book, amount to a stack that were considerably higher than the diameter of our observable universe. Thus a universe filled with black holes is much more likely than ours. Yet we don't observe such a black hole entropy dominated universe – and we couldn't even live in one. Viewed in this light, $1{:}10^{10^{122}}$ becomes a requirement for our existence.

One might argue therefore on the basis of the weak anthropic principle [13, 153] that we should not wonder about the low entropy, because if it were much higher, we could not exist and there would be no one to wonder about it. So low overall entropy is certainly a precondition for complex life. However, a much higher overall entropy would suffice, making such an argument very unconvincing. Therefore the anthropic principle is insufficient for a comprehension of time's direction, because the observable universe is much more ordered than would have been necessary for human existence. To be more accurate: the probability of our entire solar system including earth and all its life-forms popping out of coincidentally fittingly arranged particles might only be $1{:}10^{10^{85}}$ – but this is overwhelmingly more probable than the $1{:}10^{10^{122}}$ for the entire observable universe. So the anthropic principle is not helpful here: neither as a mere tautology stating a necessary condition for life nor as a selection criterion for a universe that makes life possible within a multiversal realm of possibilities, because even if there were $1{:}10^{10^{122}}$ universes differing in their initial conditions, this would not render the actual value of entropy in our universe plausible.

6 Beyond the Big Bang

We exist in a world full of order that is friendly to life in the thermodynamical sense because the big bang was supremely "orderly". And, as most scientists are convinced by now, this is exactly the reason why the universe runs like a clock – indicating a clear direction of time. But what was it that wound up the cosmic clockwork? How did this supremely special big bang come about? What caused the low entropy of the early universe?

Some 13.7 billion years ago the observable universe evolved from an extremely hot and dense region smaller than an atom which expanded enormously. While the aftermath of this big bang is both theoretically and empirically well established, and to a large extent understood, it is still a mystery as to how and why the big bang occurred at all. Was it the beginning of space and time, or only of matter? If it was a transition, what came before? If not, how could "everything" appear out of "nothing"? And was it a singular event or one of perhaps infinitely many. Do other universes also exist, and did they or will they interact with our own? These are difficult questions and controversial issues – but no longer beyond the scope of science. In modern quantum cosmology a lot of competing scenarios are being pursued [167]. They open up the exciting prospect of going "beyond" the big bang and even of finding a physical explanation for it.

"Ad fontes" ("to the sources") – this humanist slogan from the early modern period could be fitting for today's physicists too: in order to understand the direction of time, they also have to discover the origin of time. Yet the early days of the universe appear as incomplete, misleading, and dark, as historical sources often are. Considering the far longer periods of time, it is surprising that anything at all should still be preserved – and that cosmologists can partly decipher it. Indeed, the observable universe might have "forgotten" much of the information it held in its primordial times. This could be a result of cosmic inflation (insofar as it actually happened). The result of this huge expansion of space is that hardly anything remains in the observable universe from the time of inflation – if inflation had a beginning at all (and has not been going on since all eternity), something that most cosmologists presume indeed. But even in this case, our universe might have separated from the inflationary epoch at a randomly late point. Less than a hundred volume doublings would have been sufficient to cover all tracks from the time before inflation. Cosmic inflation has even been assumed to be the source of the low entropy of our universe [4]. Yet it seems that inflation alone could not have accomplished all of this (e.g., [100]). On the other hand, it might at least be the key to the door of such a deeper explanation that would have to make the initial conditions of inflation understandable – something that can be criticised as yet another shift of the problem however.

In the end, the breakthrough to a deeper understanding will be up to the theoreticians – in the form of a theory of quantum gravity that would have to be confirmed howsoever. The challenges are enormous, and the consequences are as yet unclear. Even our old companion time will probably not be left unblemished. It seems to dissolve entirely in the noise of the smallest scales of nature where there are no longer any clear, regular oscillations, and hence also no "clocks" (see [84]). The disturbing consequences of the theory of relativity – which reduced time to a "fourth dimension" and merged it with space into a unity [115] – cannot be reversed in a quantum theory of gravity, but are here to stay. General relativity implies that there is no background spacetime – no stage where things move autonomically, without affecting spacetime. Hence, there is no "time" that everything could flow along. This seems to be even more true for a theory of quantum gravity. Here the notion of a spacetime continuum breaks down at the Planck scale, turning lengths and time intervals into quasi-discrete entities. Perhaps the world must be described without a concept of time on its fundamental level [83, 123].

Nevertheless the big bang still appears special, and the arrows of time, whether fundamental or not, deserve an explanation. This might even reach beyond the big bang. Actually the big bang was not necessarily the absolute beginning of everything. Whether it was or whether it happened, on the contrary, as a phase transition – for example a "bounce" of an earlier, contracting universe or an accidental fluctuation within a quantum vacuum – is an open and very controversial issue (see below). But in principle the fluctuation or bounce scenario is a promising candidate for a dynamic origin and, thus, explanation, of the arrow of time.

Furthermore, if the big bang was not the beginning of everything but a phase transition, one need not ask how something came out of nothing (which is of course a different question than why there is something rather than nothing): the big bang then was not something that sprang into existence ex nihilo, and nor did spacetime or energy or the laws of nature. It is also meaningless to ask why the entropy was so small at the beginning, if there was no ultimate beginning at all. Nevertheless this question reappears in a modified form: Why was the entropy so small at the bounce or at the beginning of the fluctuation? If it had been large, the big bang would not have produced the smooth, low-entropy universe which is still observed today, but a chaotic mess.

7 Big Fluctuation

If the big bang was a fluctuation, the special low entropy of the universe could have originated as a pure accident (and therefore could be explained away).

But even if our observable universe were only a coincidentally developed island of order in a much greater ocean of chaos – a statistical fluctuation, as Ludwig Boltzmann deliberated as early as 1895 – then it would still be incomprehensible why this fluctuation is so persistent (Fig. 2). After all, about 13.7 billion years have passed since the big bang. But it appears to be much more probable for the spontaneous fluctuation to have arisen only last Thursday or a few seconds before this very moment right now – with all the pseudo-traces of an alleged past: the memories of earlier tax declarations and children's birthdays, the fossils of dinosaurs, the meteorites from the beginning of the solar system and the cosmic background radiation from the aftermath of the big bang itself. In a nutshell: such a bogus-universe – or only a single brain in which such a pseudo-world manifests itself – should arise overwhelmingly more frequently simply by chance than a highly structured, ordered space of at least 100 billion light years in diameter. This often disregarded objection was already made (roughly) by Carl Friedrich von Weizsäcker [176] in 1939 and reappeared in modern dark energy cosmology as the problem of the Boltzmann brains [39, 90, 161]. To give some thermodynamical numbers of entropy fluctuations in a de Sitter background, the probability of our observable universe, $1{:}10^{10^{122}}$, is extremely tiny in contrast to a spontaneous ex nihilo origination of a freak observer, perhaps $1{:}10^{10^{21}}$ for the smallest possible conscious computer, and between $1{:}10^{10^{51}}$ and $1{:}10^{10^{70}}$ for a "Boltzmann brain" [36]. (Thus, there is a controversial discussion going on about wrong assumptions underlying those kinds of estimates – not because many scientists believe that such a solipsistic illusion is true, but because these probabilities indicate possible errors in cosmological reasoning and deep difficulties of multiverse models, especially the measure problem in inflationary cosmology.)

Of course, a virtue can be made out of necessity. Sean Carroll and Jennifer Chen [30] did just that. They argue that our universe really is a mere fluctuation

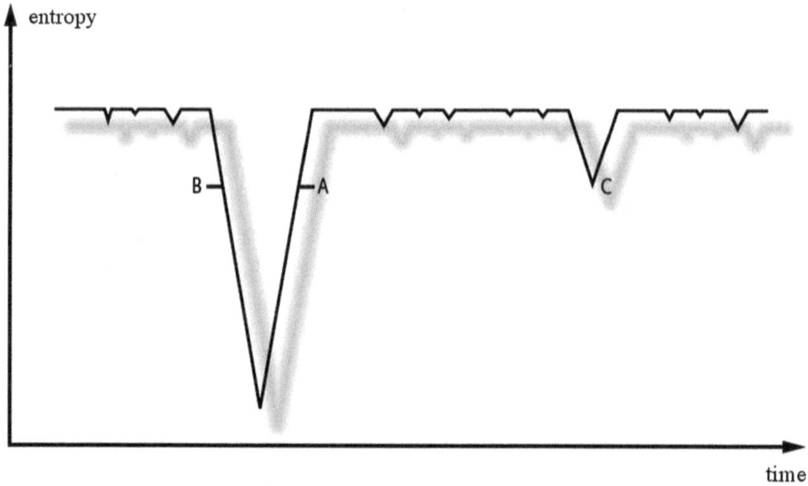

Fig. 2 Order from chaos: When a system is in a state of maximum disorder – i.e., entropy – then temporary "islands of order" and thus local directions of time (point C) develop by means of chance processes over long periods of time. That's why there have been recurring speculations about the entire observable universe being such an island in the midst of chaos. Intelligent observers could only live within one of these "entropy gradients" (point A). Yet there are two fundamental difficulties with such a viewpoint. First, it would be much more likely for everything around A to have originated out of chaos only very recently (as in the case of C) – but then most of what seems to have happened in the past would be a mere illusion. Second, life-forms in the vicinity of B would experience the direction of time exactly in reverse to A.

among myriads. This is possible if the entirety of empty space, taken as a quantum vacuum, contains even more entropy than isolated black holes that only have the maximum entropy within a specific volume. Such an (eternal) accelerating expansion of space, driven by the still mysterious dark energy, could indeed entail an even higher entropy than black holes. Yet such a vacuum must produce random quantum fluctuations again and again. Some of them become huge because of inflation, until they deflate entirely due to the perpetual expansion caused by dark energy. And such cycles, according to Carroll and Chen, are more likely than random fluctuations of dinosaurs and bogus-universes. Thus, in an infinite future, time might not be a problem. Eventually, anything could spontaneously pop into existence due to quantum fluctuations if spacetime is eternal. They would mostly result in meaningless garbage, but a vanishingly small proportion would contain people, planets, and parades of galaxies. This book will also reappear again (a modern version of the well-known philosophy of eternal recurrence, which has many other versions in current cosmology too, see [168]). And this kind of quantum resurrection might even spark a new big bang. According to Carroll and Chen, one must be patient, however, and wait some $10^{10^{56}}$ years for another recurrence of our observable universe (if a de Sitter vacuum with a positive cosmological constant Λ is the "natural ground state"). Our whole universe might be such an island in a Λ-sea, i.e., a dark energy dominated vacuum (see also [37, 38] for a different, but related scenario). Still, there is no preferred direction of time in this scenario, because in both directions of time, baby universes emerge by means of fluctuations, empty out, and beget babies of their own. On an extremely large scale, such a multiverse looks time-symmetric on average – in the "past" as well as in the "future", new universes come into existence and reproduce without limit. An arrow of time appertains to each one of them, yet in half of the cases it points oppositely to the others.

If it is believed that such cycles are bogus as well, one has to look somewhere else for an explanation of the low initial entropy of space. In other words, if the stream of time is not a coincidence, then it must flow from a source. And this might provide a selection of initial conditions. Perhaps our universe is indeed the result of a natural selection process within a multiverse, such as the string landscape [142], if many newborn universes rapidly vanish. Laura Mersini-Houghton [102] proposes that the out-of-equilibrium quantum dynamics of the landscape of the initial patches may have provided such a selection mechanism. The tendency of any system is to reach equilibrium by maximizing entropy. But as matter degrees of freedom tend to equilibrium by trying to pull the initial patch to a black hole crunch, the gravitational degrees of freedom contained in the vacuum energy tend towards equilibrium by trying to expand that initial patch to infinity. These opposing tendencies in this tug-of-war on the initial patch guarantee that this system is far from equilibrium, and that its dynamics selects the initial conditions of the universe. If the vacuum energy wins over matter, the initial patch grows, giving rise to an inflating survivor universe. But if the pull of matter is stronger, then the initial patch cannot grow and contracts, resulting in a terminal universe. Therefore, in the ensemble of all possible initial conditions and energies on the string landscape multiverse, where every vacuum hosts a potential birthplace for a universe, only a fraction of them are selected as survivor universes, namely initial states with high vacuum energies. Although an ensemble of all possible initial states has shrunk to a subset by being wiped clean of the low energy patches, it still contains a whole multiverse of survivor universes. Such a selection process might also be a key to understanding the arrow of time because survivor universes start at high energies and therefore low entropies. Thus the direction of time would have a dynamical origin. And if the ergodicity of the phase space is broken, a universe cannot fluctuate back close enough to its previous state. If so, dynamics would not allow temporary or eternal duplications. According to Mersini-Houghton our cosmic domain is connected with everything else in the multiverse through nonlocal quantum entanglement, inherited from the entanglement of the initial patches, which cannot be lost. And these superhorizon-sized connections left imprints on the cosmic microwave background and large scale structure. Actually she predicted the existence of two giant voids (one might have been detected already [97, 127, 159]), a dark flow of galaxies (some indications of which have also been discovered [81, 165]), a higher supersymmetry breaking scale (which could be tested by particle colliders like the Large Hadron collider) if our universe acquired its initial vacuum energy by breaking supersymmetry, and certain cosmic microwave background features (a suppressed amplitude of perturbations but an enhanced power with distinct signatures at higher multipoles in the power spectrum, which is in agreement with recent measurements).

8 Big Bounce

Another possibility is that the big bang originated out of a big bounce, a collapse of a precursor universe or spacetime [104]. To explain the arrow of time, the bounce had to be special. There are basically two possibilities: either entropy rises forever, constituting a persistent arrow of time through all eternity, or entropy decreases during the contracting phase before the bounce and increases after it.

If entropy rises forever and the bounce is the end of an infinitely long contraction, from eternal past to the moment of the bounce, the entropy somehow stayed quite low, or else it was reduced at the bounce at least in a region that became the observable part of the universe. Perhaps there was a "cosmic forgetfullness" at work, destroying entropy and/or information from the precursor universe (see [17,19,20]).

It is also possible that, while global entropy increases, the entropy density within local horizons or Hubble volumes decreases from one cycle to the next due to the dark-energy-driven exponential expansion (see [87,88,140]). Furthermore, there are scenarios – relying on a strange form of dark energy called phantom energy – that describe entropy decrease before/during the bounce [27,50]. Whether our universe has such subtle properties is, however, questionable and awaits substantiation.

Alternatively, entropy must have had a minimum value at the bounce and was in fact higher on the other "side", whence it increased both in the contracting phase before the bounce and the expanding phase after it. If so, the evolution of the universe, both in size and in entropy, would be symmetric in time. The time-reversal symmetry of the laws of physics would be accompanied by the large-scale behavior of the universe. And the pitfall of temporal chauvinism [116] – the temptation to treat the "initial" state of the universe differently from the "final" state – would also have been avoided. Very recently indeed some bounce models came up with higher entropy and thus opposite thermodynamic arrows of time on either side of the bounce: both in Euclidean quantum gravity ([69,70,108]; see also [169]) and in loop quantum gravity [19,20] (for opposite arrows in an inflationary scenario see [1,2]).

The cost of this advantage is a new problem however: Why was the entropy so low at the bounce? So in a time-symmetric scenario the mystery lies not in the infinite or most distant past beyond the big bang, but right in the temporal "center" of the universe (assuming it will expand forever). Is the fine-tuning problem therefore just shifted again – shifted back to the big bang, even if it were not an inexplicable singularity? And if entropy is increasing away from the bounce in both directions, one can argue that there are two opposite arrows of time which prevent a causal explanation of the bounce. Instead, the bounce can then be understood as the (uncaused?) origin of two unrelated universes. This would explain away the problem of time's direction, because both directions are realized; thus globally there is no fundamental asymmetry. But the mystery of the low entropy at the beginning of the universe(s) would still remain.

9 A Fundamental Asymmetry?

A straightforward explanation for the arrow(s) of time would be a fundamental derivation of a law or genuine property of nature for understanding the low entropy as an initial condition with a fundamental origin, not a dynamic one. This basic asymmetry might be encoded within the spacetime structure itself, i.e., the arrow of time would be an intrinsic property of spacetime and does not need to be reduced to non-temporal features [40].

Matias Aiello, Mario Castagnino and Olimpia Lombardi [3] for instance argue that the subset of time-symmetric spacetimes has measure zero in (or is a proper subspace of) the set of all possible spacetimes admissible in general relativity, because symmetry is a very specific property, whereas asymmetry is highly generic. Therefore, in the collection of all physically possible spacetimes (if it exists!), those endowed with a global and non-entropic arrow of time should be overwhelmingly probable, whereas non-existence of the arrow of time would require an extraordinarily fine-tuning of all the variables of the universe. Furthermore, the global time-asymmetry could be transferred to local contexts as an energy flow pointing to the same temporal direction over the whole of its spacetime. Thus in this approach the arrow of time is defined by the time-asymmetry of spacetime and expressed by the energy flow.

Brett McInnes [99, 100] speculates about a geometrical explanation. Due to the homogeneity of the cosmic microwave background, low entropy is more precisely low gravitational entropy and should be based upon spatial uniformity. Thus it has an intrinsically geometric form and root. McInnes therefore argues that the initial geometry must have been smooth (a perfectly flat spacelike surface, which was exactly locally isotropic around each point) and that there was a beginning of time, because something which has no past cannot be distorted by any "prior" conditions. Furthermore, baby universes can only have an arrow of time if they inherit one; if so, the problem of explaining the arrow is reduced to its explanation in the case of the "first" universe which would have come out of "nothing". If it originated along a spacelike surface with the topology of a torus, recent results in global differential geometry suggest that the geometry of this surface had to be non-generic. This geometric specialness would be communicated then to matter through the inflaton. The first universe inflated and gave birth to baby universes which inherited the arrow of time.

John Richard Gott III and Li-Xin Li [57, 58], cf. [166]) suggested another explanation for the arrow of time. Their model of a self-creating universe assumes a cold, low entropy time loop at the onset of the multiverse, a curled-up small time dimension, comparable to the compactified extra dimensions in string theory. In this model a region of a closed timelike curve allows the universe to be its own "mother". Thus the universe would not have been created out of nothing but out of something – itself. This also avoids the problems associated with both an eternal spacetime and an absolute beginning. Although it is sometimes said that asking what happened before the big bang is meaningless, like asking what is south of the South Pole, by supposing the universe (or the first of a multiverse) broke out of a time loop, asking what was the earliest point might be like asking what is the easternmost point on the Earth: one can keep going east around and around the Earth – there is no easternmost point. For every event in the very early universe there would have been events that preceded it, and yet the universe would not have existed eternally in the past. This also offers a new kind of cosmology in respect to time as the class of pseudo-beginning models do (see below).

10 A Universe With or Without a Beginning

Did the universe have a beginning or does it exist forever, i.e., is it eternal at least in relation to the past? This fundamental question was a major topic in ancient philosophy of nature and the Middle Ages. Philosophically, it was then more or less banished by Immanuel Kant's "Critique of Pure Reason" [80]. He argued that it is possible to prove both that the world has a beginning and that it is eternal (first antinomy of pure reason, A 426f/B 454f). Kant believed he could overcome this "self-contradiction of reason" ("Widerspruch der Vernunft mit ihr selbst", A 740) by what he called "transcendental idealism". After that the question as to whether the cosmos exists forever went out of fashion in philosophical discussions. This is somewhat surprising, because Kant's argument is quite problematic (see, e.g., [48, 75, 79, 93, 130, 135, 178, 179]). In the 20th century, however, the question once again became vital in the context of natural science, culminating in the controversy between big bang and steady state models in modern physical cosmology ([86]). In recent years, it has reappeared in the framework of quantum cosmology [147, 148], where, on the one hand, there are models that assume an absolute beginning of time while other scenarios suppose that the big bang of our universe was only a transition from an earlier state, and that there are perhaps infinitely many such events.

General relativity breaks down at very small spatiotemporal scales and high energy densities, leading to singularities. This is why quantum cosmology is needed. But in contrast to the framework of general relativity, which is theoretically well understood and has been marvelously confirmed by observation and experiment, the current approaches in quantum cosmology are still quite speculative, controversial, and almost without any empirical footing as yet. Nevertheless they offer promising new prospects – not only for explaining the big bang, but also for solving the problems of time.

Note that "big bang" is an ambiguous term which has led to some misunderstandings and prejudices. One should draw a distinction between at least four logically different meanings [154, 163]:

(1) the hot, dense early phase of our universe where the light elements were formed,
(2) the initial singularity,
(3) an absolute beginning of space, time, and energy, and
(4) only the beginning of our universe, i.e., its elementary particles, energy, vacuum state, and perhaps its (local) spacetime, its arrow of time, its invariants (described by its laws and constants of nature).

That our universe originated from a big bang in the meaning of (1) is almost uncontroversial, while (2) is relativistic cosmology's limit of backward extrapolation where the known laws of physics break down. Different models of quantum and string cosmology try to overcome this limit, and (3) and (4) are two distinct possibilities, broadly classifying many different competing models. Those characterized by (3) are initial cosmologies. They postulate a very first moment (see [59], [136]) or a limit or boundary of the past. Those characterized by (4) are eternal cosmologies. There are different kinds of them, in both ancient and modern cosmology: static ones (without irreversible changes on a coarse-grained level), evolutionary ones (with cumulative change), and revolutionary ones (with sharp phase transitions). They could have either a linear or a cyclic time. Option (4) also allows the possibility that there are other universes (for different notions of "universe" and "multiverse" see, e.g., [154, 163]) and that our universe neither exists eternally, nor came into being out of nothing or out of a timeless state, but that space and time are not fundamental and irreducible at all, or that there was a time "before" the big bang – "big bang" in the sense of (1). Such pseudo-beginning models offer (as time-loop models do) a third option between initial and eternal cosmologies. Thus, although there is already an overwhelming plentitude of cosmological models [167], from a conceptual point of view there are not many options. Insofar as they try to explain the big bang they can be classified roughly into just four types with respect to time (Table 1). Most of them might be realized either uniquely as one universe, or as a multiverse, especially if "spatial branching" is possible – as, for instance, in eternal inflation.

Both initial and eternal cosmologies involve severe explanatory problems, so it is useful to search for alternatives.

Eternal cosmologies need not assume a first cause or accident, but they shift the burden of explanation into the infinite past. Although every event might be explicable by earlier events and causal laws, eternal cosmologies cannot even address the questions as to why a temporally infinite cosmos exists and why it is the way it is. And there might be even deeper problems. Since we are able to assign a symbol to represent "infinity" and can manipulate such a symbol according to specified rules, one might assume that corresponding infinite entities (e.g., number of particles or universes) might exist. But the actual (i.e., realized in contrast to potential or conceptual) physical (in contrast to mathematical) infinity has been vehemently criticized as not being constructible, implying contradictions (see [76]

Table 1 Big bang cosmologies with respect to time

types of cosmology	sub-types	universal	multiversal
initial: absolute beginning		yes	yes
eternal: no beginning	• steady state	yes	no
	• quasi-steady state	yes	yes
	• bounce (with/out arrow of time reversal)	yes	perhaps
	• oscillation (cyclic)	yes	yes
time-loop	• at the beginning	yes	yes
	• the universe as a whole	yes	yes
pseudo-beginning	• from a static state	yes	perhaps
	• from a fluctuating vacuum	no	yes

and [141], ch. 5). If this were correct, it should also apply to an infinite past. (A future-eternal cosmos might be less problematic, if it is viewed as an unfolding, unbounded, i.e., only potential one.) This is a controversial issue, but it might be seen at least as another motivation to search for alternatives to past-eternal cosmologies.

Initial cosmologies, on the other hand, run into deep metaphysical troubles to explain how something could come out of nothing and why is there something rather than nothing at all (see [105, 157]). Even the theological doctrine of creatio ex nihilo does not start with nothing at all but with something, that is God, so the principle "ex nihilo nihil fit" still holds. And contemporary secularized ex-nihilo initial cosmologies usually claim, as Alexander Vilenkin has said (quoted in [150] (p. 45); cf. [175] (p. 205)), that there were at least the laws of physics, even if there was nothing else. (Concerning his own model, Vilenkin [174] (p. 26) admitted that "the concept of the universe being created from nothing is a crazy one", and an analogy with particle pair creation only deepens the problem, because matter-antimatter particles do not pop out of nothing, but are transformations of energy which is already there.) Similarly, Heinz Pagels [109] (p. 347) subscribed to some kind of platonism with respect to physical laws: "This unthinkable void converts itself into the plenum of existence – a necessary consequence of physical laws. Where are these laws written into that void? What "tells" the void that it is pregnant with a possible universe? It would seem that even the void is subject to law, a logic that exists prior to space and time." And Stephen Hawking [73] (p. 174) asked: "Even if there is only one possible unified theory, it is just a set of rules and equations. What is it that breathes fire into the equations and makes a universe for them to describe? The usual approach of science of constructing a mathematical model cannot answer the question of why there should be a universe for the model to describe. Why does the universe go to all the bother of existing?" But if one does not subscribe to an origin of something (or everything) from what is really nothing, one need not accept platonism with respect to the ontological status of physical laws [150]. They might simply be seen as the outcome of invariant properties of nature. If so, they do not govern nature but are instantiated from it. They are abstract descriptions – just as a model or theory of reality is not to be confused with reality itself.

So from what has been sketched here, it is helpful to search for a "third way" between initial and eternal cosmologies to explain more than the latter but not fall into the problems of the former. Pseudo-beginning cosmologies, as they can be called, might offer such a middle course [154, 156, 162]. Based on a distinction between two kinds of time scale, microscopic and macroscopic, these models also offer a conceptual and perhaps physical solution of the temporal aspect of Kant's "first antinomy of pure reason", i.e., showing how our universe could in some sense have both a beginning and an eternal existence.

11 Microtime, Macrotime and the Pseudo-Beginning Proposal

Kant's first antinomy makes the error of the excluded third option, i.e., it is not impossible that the universe could have both a beginning and an eternal past. If some kind of metaphysical realism is true, including an observer-independent and relational time, then a solution of the antinomy is conceivable [154]. It is based on the conceptual distinction between a microscopic and a macroscopic time scale.

Only the macroscopic scale is characterized by an asymmetry of nature under a reversal of time, i.e., the property of having a global or at least wide-ranging evolution – an arrow of time – or many arrows, if they are independent from each other. Thus, the macroscopic (coarse-grained) scale, or macrotime for short, is by definition temporally directed – otherwise it would not exist. (It shall not be discussed here whether such an arrow must be observable in principle, which would raise difficult questions, e.g., in relation to an empty, but globally expanding universe.)

On the microscopic scale, however, there exist only local, randomly distributed events without dynamical trends, i.e., without a wide-ranging time-evolution or an increase of entropy density. This time symmetry occurs if one or both of the following conditions are satisfied: first, if the system is in thermodynamic equilibrium (e.g., if there is a huge number – or degeneracy – of microscopic states identifiable with the same coarse-grained macroscopic state); and/or, second, if the system is in an extremely simple ground state or metastable state. (Metastable states are local, but not global minima of a potential landscape and hence can decay; ground states might also change due to quantum uncertainty, i.e., due to local tunneling events or fluctuations.) Thus, while microtime always exists, provided there are at least some quantum fluctuations, macrotime could vanish or may be absent altogether. In conclusion, conceptually there are two kinds of timelessness: genuine timelessness, which means the absence of both micro- and macrotime; and effective timelessness, in which only macrotime is absent.

Distinguishing between micro- and macrotime does not rely on whether – and in which sense – time is real or not. It is therefore advantageous or effective to use temporal concepts even if time is ultimately taken to be an illusion (for example because only spacetime is real, because time is not fundamental, or because it emerges only within consciousness or as a fiction of language). Hence the distinction of micro- and macrotime remains helpful and is largely independent of ontological considerations.

A system with microtime but no macrotime does not need to have massive particles; events of any kind, e.g. quantum fluctuations or gravitational wave interference would suffice. One might argue that for any kind of time there must be at least some processes which could, in principle, serve as a clock. If so, then any kind of microscopic oscillation would be enough for the persistence of microtime. If photons are exchanged or scattered, for instance, then there is clearly change – at least conceptually. This is sufficient for the existence of microtime – although not for a sophisticated clock, which would not only depend on periodic processes but would also need to be able to track and record them. Such storage ability, or "memory", requires a (local) thermodynamic non-equilibrium and thus macrotime.

If macrotime is not fundamental, the arrow(s) of time could have a beginning – and therefore an explanation. Some still speculative theories of quantum gravity permit the possibility of a global, macroscopically timeless ground state (e.g., quantum or string vacuum, spin networks, twistors). Due to accidental fluctuations, which exceed a certain threshold value, universes might emerge out of that state. Due to some also speculative physical mechanism (like cosmic inflation) they acquire – and thus are characterized by – directed non-equilibrium dynamics, specific initial conditions, and hence an arrow of time. (It could be defined, for instance, by the cosmic expansion parameter or by the increase of entropy.) Note that, strictly speaking, such universes are not "inside" or "embedded in" the vacuum ground state, but cut their cords and exist in some respects "somewhere else".

Systems with an arrow of time undergo a directed development. This is manifest only on a macroscopic scale. Such a macroscopic time, or macrotime for short, comes along with an increase of entropy. This macroscopic global time-direction is the main ingredient of Kant's first antinomy, for the question is whether this arrow has a beginning or not. To get a simplified idea of macro- and microtime, classical thermodynamics and statistical mechanics can serve as an analogy (Figs. 3 and 4, see [4]).

For example molecules in a closed box (Fig. 3) spread from a corner (1) – if they were released there, for instance, from a gas cylinder – in every direction and eventually occupy the whole space (3). Then a state of equilibrium is reached which has no directed development anymore and thus no macrotime. Coarse-grained "low-resolution snapshots" of the whole system or sufficiently large parts of it show no difference (3 and 4). On a fine-grained level there are still changes (3 versus 4). Thus, a microtime always remains. Due to accidental, sufficiently large fluctuations – which happen statistically even in a state of equilibrium if there is enough microtime

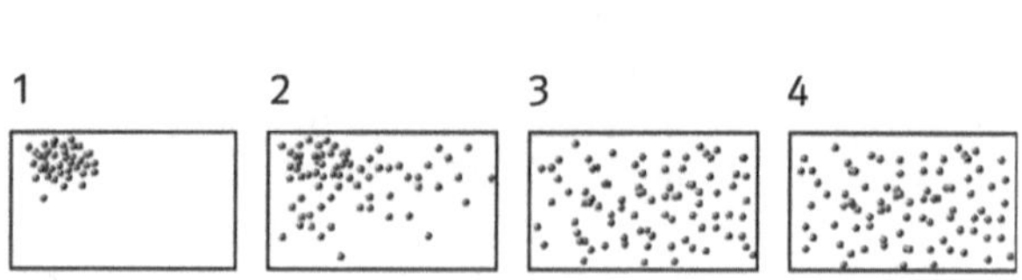

Fig. 3 Micro- and macrotime in a closed system.

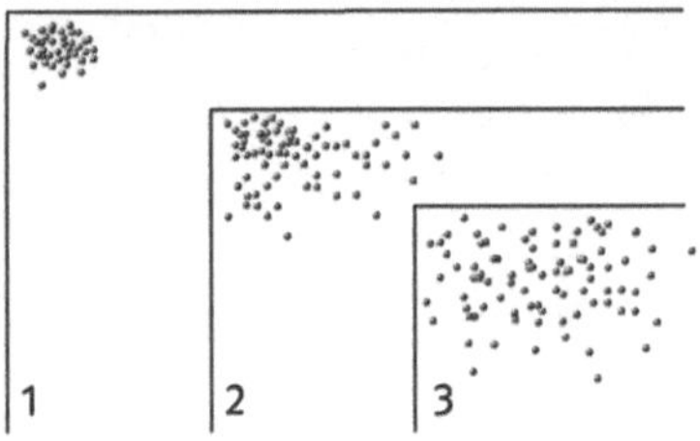

Fig. 4 Micro- and macrotime in an open system.

available – local structures can arise (from 3 to 1) and a macrotime temporarily comes into being again.

If the system is not closed but open (Fig. 4), a state of equilibrium does not necessarily develop, and macrotime does not vanish. For instance, in the universe this is the case because space expands. Whether there were specific, improbable initial conditions at the big bang (1) or whether order and a directed development could have come out of quite different initial configurations is controversial. Possibly the whole universe is an accidental fluctuation in a macrotimeless quantum vacuum.

In conclusion, and contrary to Kant's thoughts, there are reasons to believe that it is possible, at least conceptually, that time both has a beginning – in the macroscopic sense with an arrow – and is eternal – in the microscopic notion of a steady state with statistical fluctuations.

Is there also some physical support for this proposal?

Surprisingly, quantum cosmology offers a possibility that the arrow has a beginning and that it nevertheless emerged out of an eternal state without any macroscopic direction of time. So this possible overcoming of the first antinomy is not only philosophically conceivable but is already motivated by modern physics. At least some scenarios of quantum cosmology, loop quantum gravity (quantum geometry), and string cosmology can be interpreted as examples for such a local beginning of our macroscopic time out of a state with microscopic time, but with a quasi-eternal, global macroscopic timelessness.

To put it in a more general, but abstract framework and get a sketchy illustration, consider Fig. 5. Physical dynamics can be described using "potential landscapes" of fields. For simplicity, only the variable potential (or energy density) of a single field is shown here. To illustrate the dynamics, one can imagine a ball moving along the potential landscape. Depressions stand for states which are stable, at least temporarily. Due to quantum effects, the ball can "jump over" or "tunnel through" the hills. The deepest depression represents the ground state.

Usually it is assumed that the state of the universe – the product of all its matter and energy fields, roughly speaking – evolves out of a metastable "false vacuum" (Fig. 5, top left) into a "true vacuum" (maybe the "ground state") which has a state of lower energy (potential). There might exist many (perhaps even infinitely many) true vacua that would correspond to universes with different constants or laws of

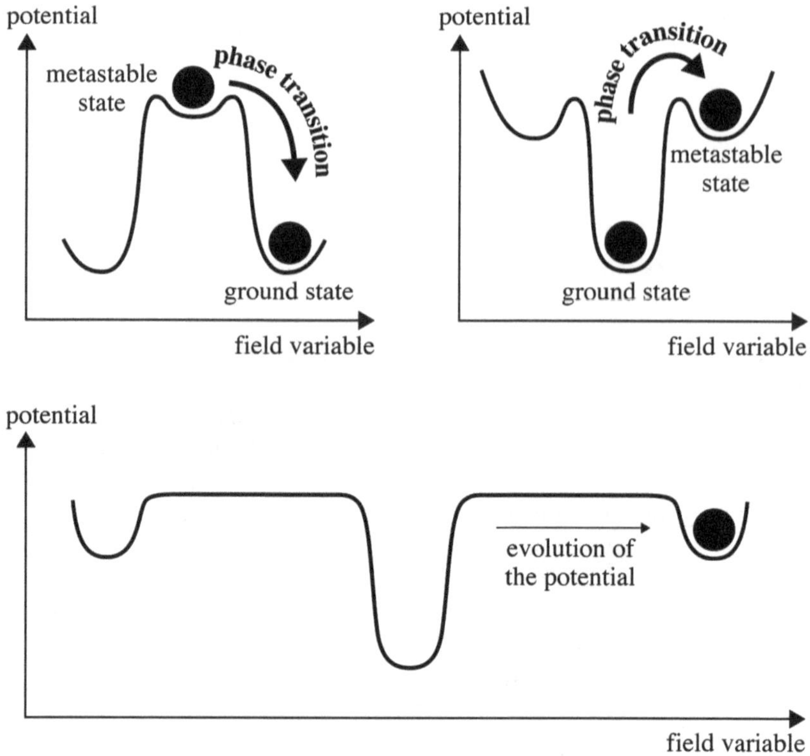

Fig. 5 Something out of almost nothing.

nature (imagine the graph in Fig. 5 not as a line but as a three-dimensional object shaped like a sombrero, where every spot on its brim corresponds to a different universe with a different set of physical laws and constants). The initial condition of this scenario, however, remains unexplainable and unlikely.

It is more plausible to start with a ground state that is the minimum of what can exist physically (Fig. 5, top right). According to this view, an absolute nothingness is impossible. There is something rather than nothing, because something cannot come out of absolutely nothing, and something does obviously exist. Thus, something can only change, and this change might be described by physical laws. Hence, the ground state is almost "nothing", but can become thoroughly "something" (more). (Therefore, it is only marginally, but qualitatively incorrect to say that there is something rather than nothing because nothingness is unstable – a difference which makes all the difference in the world and, in fact, makes the whole world.) Possibly, our universe – and, independent from this, many others, probably most of them having different physical properties – arose from such a phase transition out of a quasi-atemporal quantum vacuum (and perhaps got completely disconnected). Tunneling back might be prevented by a deformation of the potential (sketched in Fig. 5, bottom). For example this might happen due to the exponential expansion of this brand new space. Because of this cosmic inflation the universe may have become gigantic, while the potential hill may simultaneously have broadened enormously and become (almost) impassable. This preserves the universe from relapsing into non-existence. On the other hand, if there is no physical mechanism to prevent the tunneling-back or make it at least very improbable, there is still another option: if infinitely many universes originated, some of them could be long-lived only for statistical reasons. But this possibility seems to be less predictive and therefore an inferior kind of explanation for not tunneling back.

An example of such a dynamical resolution of the arrow of time problem was recently sketched by Raphael Bousso [24] (see also [143]). He has shown how, in principle, the string theory landscape can give rise to an arrow of time, independently of the initial entropy and without creating a Boltzmann brain problem. If certain assumptions hold, upward fluctuations spontaneously create metastable vacua in which the expected number of ordinary observers is greater than that of Boltzmann brains. If this proposal is correct, or an analogous one within a different theoretical framework, low-entropy boundary conditions are neither necessary nor sufficient for macrotime to arise.

The micro-/macrotime distinction is neutral in respect of whether opposite arrows of time could emerge from a bouncing region or a fluctuating vacuum. This issue is model-dependent (provided that time's direction is not fundamental). No temporal chauvinism should be assumed a priori, but this only poses the question and does not answer it. Nor does it solve the problem whether and how one can know about a global time-symmetric cosmos. Opposite arrows arising from a bounce at least were in some contact, though it is doubtful how influences could propagate to the other "side" of time, if the causal arrow comes along with the other arrows. Opposite arrows arising from a huge vacuum, however, could be arbitrarily far from each other. Perhaps quantum entanglements left their mark, but how might they keep track of time's direction in other universes?

Another crucial question remains even if universes could come into being out of fluctuations of (or in) a primitive substrate, i.e., some patterns of superposition of fields with local overdensities of energy: Is spacetime part of this primordial stuff or is it also a product of it? Or, more specifically, does such a primordial quantum vacuum have a semi-classical spacetime structure or is it made up of more fundamental entities?

Both alternatives have already been investigated in some respect. Unique-universe accounts, especially the modified Eddington models – the soft bang/emergent universe – presuppose a kind of semi-classical spacetime [14,43,44,121]. The same is true for some multiverse accounts describing our universe (and others) as a fluctuation or collapse/tunnel event – e.g., the models explained in [11,25,26,30,37,38,49,56,77,138,146] – where Minkowski space, a tiny closed, finite space or the infinite de Sitter space is assumed. The same goes for string theory inspired models like the pre-big bang scenario of [53, 151, 173], because string and M-theory is still formulated in a background-dependent way, i.e., requires the existence of a semi-classical spacetime (for some speculative ideas in string theory that go beyond this, i.e., treating spacetime as emergent, see, e.g., [34]).

A different approach is the assumption of "building blocks" of spacetime, a kind of pregeometry [110], and also the twistor approach (see [74], ch. 6), and the cellular automata approach [180]. The most elaborated accounts in this line of reasoning are loop quantum gravity ([7, 10, 19, 123, 124, 126]; see also [152], and for comparisons with string theory [68, 110, 126]) and its relatives such as spin foams or dynamical triangulation [5]. Here, "atoms of space and time" underlie everything. According to the theory of loop quantum gravity, for example, spin networks made out of one-dimensional structures are the "building blocks" of reality [10]. Their connections and excited states define matter and forces, but also lead to the emergence of spacetime in the first place. Thus, spacetime would not be a foundation of nature but a subordinate product – in the end an illusion, "albeit a very stubborn illusion", as Albert Einstein already assumed in the context of relativity. Abhay Ashtekar likes to quote Vladimir Nabokov in this context: "Space is a swarming in the eyes, and Time a singing in the ears". Carlo Rovelli [123, 125] also assumes time to be a phantasmagoria.

Though the question whether semiclassical spacetime is fundamental or not is crucial, an answer might nonetheless be neutral with respect to the micro-/macrotime distinction. In both kinds of conceptualizations of the quantum vacuum, the macroscopic time scale is not present. And the microscopic time scale has to be there in some respect, because fluctuations represent change (or are manifestations of change). This change, reversible and relationally conceived, does not occur "within" microtime, but actually constitutes it. Out of a total stasis (even on the microscopic level) nothing new and different can emerge, because an uncertainty principle – fundamental for all quantum fluctuations – would not be realized. In an almost, but not completely static quantum vacuum, however, nothing changes macroscopically either, but there are microscopic fluctuations.

Surely a background-independent approach is more promising philosophically, for it is simpler and more fundamental and takes the lesson of general relativity more seriously. The concept of microtime can nevertheless be applied to such a "spacetime dust". A good illustration (or candidate) is the spin foam approach in loop quantum gravity (see [9, 106, 114]), but in principle spin networks would also suffice (see [123, 124]), and some versions of loop quantum cosmology are indeed related to a pseudo-beginning [8, 18, 19].

To summarize, the concept of a pseudo-beginning of our universe (and probably of infinitely many others) is a viable alternative to both initial and past-eternal cosmologies. Note that this kind of solution bears some resemblance to a possibility of avoiding the spatial part of Kant's first antinomy, i.e., his claimed proof of both an infinite space without limits and a finite, limited space: the theory of general relativity describes what was considered logically inconceivable before, namely that there could be universes with finite, but unbounded space [42], i.e., this part of the antinomy also makes the error of the excluded third option. Thus, the pseudo-beginning proposal offers a way of steering a middle course between the Scylla of a mysterious, secularized creatio ex nihilo, and the Charybdis of an equally inexplicable eternity of the world.

If our universe has a beginning within the multiverse, one could object that it was only shown that certain parts of the world had a beginning, but not the world as a whole, i.e., the sum of all its parts. Thus, the question would repeat itself, although in a larger context. And, indeed, this is already an issue of contemporary cosmology.

For example Alvin Borde, Alan Guth and Alexander Vilenkin [22] argued that, within the framework of a future-eternal inflationary multiverse, as well as some more speculative string cosmologies, all worldlines are geodesically incomplete and, thus, the inflationary multiverse has to have a beginning. Unfortunately, if future-eternal inflation is true, all "hypotheses about the ultimate beginning of the universe would become totally divorced from any observable consequences. Since our own pocket universe would be equally likely to lie anywhere on the infinite tree of universes produced by eternal inflation, we would expect to find ourselves arbitrarily far from the beginning. The infinite inflating network would presumably approach some kind of steady state, losing all memory of how it started [...]. Thus, there would be no way of relating the properties of the ultimate origin to anything that we might observe in today's universe" ([63], p. 78). (However, there is some discussion whether early inflationary traces could remain nevertheless, and thus a cosmic persistence of memory, see, e.g., [51, 52]. It is also controversial whether eternal inflation happened at all, see [101, 139] for crucial problems.)

On the other hand, Andrei Linde and – at least for the sake of argument – Anthony Aguirre and Steven Gratton argued that the multiverse could be past-eternal nevertheless, because either all single world lines would have to start somewhere, but not the whole bundle of them [89], or there could exist some (albeit strange) spacetimes with single past-eternal world lines [1, 2].

This issue is not settled, and even in those scenarios a global arrow of time may not necessarily exist. However, other frameworks are possible – and they have already been developed to some extent – where a future-eternal inflationary multiverse is both not past-eternal and beginningless, but arises from some primordial vacuum that is macroscopically timeless. Thus, again, the beginning of some classical spacetimes is not equivalent to the beginning of everything.

One can also imagine that there is no multiverse, but that the whole (perhaps finite) universe – ours – was once in a steady state without any macroscopic arrows of time but, due to a statistical fluctuation above a certain threshold value, started to expand ([121] and, independently, [14, 43–45]) – or to contract, bounce, and expand – as a whole, and acquired an arrow of time. In such a case the above-mentioned reply, which was based on the spatial distinction of a beginning of some parts of the world and the eternity of the world as a whole, would collapse.

Nevertheless, it is necessary to distinguish between the different notions and extensions of the term "universe". In the simplest case, Kant's antinomy might be based on an ambiguity of the term "world" (i.e., the difference between "universe" and "multiverse"), but it does not need to be; and it was not assumed here that it necessarily was.

The temporal part of Kant's first antinomy was purely about the question whether the macroscopic arrow of time is past-eternal or not. And if it is not past-eternal this does not mean that time and hence the world has an absolute beginning in every respect – it is still possible that there was or is a world with some underlying microscopic time. Thus the pseudo-beginning proposal at least shows that there is a promising third option besides Kant's dichotomy and antinomy, but of course this proves neither that such a possibility is the only consistent one nor that, as a matter of fact, our universe arose from such a time before time.

12 Conformal Cyclic Cosmology

According to the conformal cyclic cosmology developed by Roger Penrose [113], the history of the universe consists of a (perhaps endless) succession of aeons, where the indefinitely expanding remote future of each aeon is in some sense identical with the big bang of the next aeon, and the entropy of the universe of the remote future seems to return to the small value that it had at the big bang. This is conceptualized as the conformal continuation of the remote future of a previous aeon of the universe; the future infinity of our universe precedes the big bang of another aeon; and this succession continues indefinitely. So the entire succession of aeons taken together provides a conformal manifold which is non-singular in the past direction.

Penrose's proposal is promising because, with spacetime conformal rescalings, the spacetime metric can be changed without affecting the light cones and thus causal relations. This is a consequence of general relativity. Nine of the ten independent components of the metric tensor $g_{\mu\nu}$ at any one point in spacetime determine where the light cone is at this point. The remaining independent component provides the scale of the metric at that point. It fixes the actual passage of time, once the causal spacetime structure has been determined, since the metric tensor is basically a measure of clock rates. The causal structure can be understood as more fundamental than the time rates because various parts of physics – for example James Clerk Maxwell's conformally invariant theory of electromagnetism – depend only on this causal or conformal aspect of $g_{\mu\nu}$, not on the scaling. This allows a kind of circumvention of the big bang singularity.

Mathematically, a cosmological singularity can be represented as a conformal boundary (hypersurface) to a spacetime, where an infinite conformal expansion is applied; and vice versa the asymptotic infinite future spacetime can be represented as a hypersurface involving an infinite conformal contraction. Furthermore, the conformal geometry of the spacetime manifold can be extended in a smooth way to a region preceding the big bang or to a region extending smoothly to beyond the future infinity. These boundaries were introduced merely as a mathematical convenience in the first place. But the conformal cyclic cosmology demands that both extended regions represent real spacetime regions. Here a past-spacelike hypersurface boundary is adjoined to a spacetime in which the conformal geometry can be mathematically extended smoothly through it, to the past side of this boundary, i.e., to a pre-big bang Lorentzian manifold. Thus the big bang singularity is expanded out infinitely into what can be described as a smooth past boundary – a spacelike hypersurface instead of a singularity, such that the conformal metric of spactime extends across that hypersurface. This transformation preserves the causal structure of the original spacetime, although it does not preserve lengths and times.

It seems confusing that the vanishing density and temperature in the infinite future can somehow be equated with an initial state with the seemingly opposite properties of an almost infinite density and temperature. And this would indeed be physically meaningless if the metric geometries of the two regions were to match. But temperature and mass density depend on the metric structure of spacetime, and cannot be determined merely from the causal (conformal) spacetime structure. In other words, matter is sensitive only to the nine components per point provided by the conformal structure of the universe, the light cones, but is blind to the single remaining component that provides the scale of the metric which would determine clock rates and distance measures.

Any particle of mass m can be seen as a "clock" which ticks away with a frequency ν that is proportional to this mass: $\nu = mc^2/h$, where h is Planck's constant. But because of the high temperature in the very early universe, the particles were effectively massless (they exceeded the Higgs mass), satisfying conformally invariant equations. If it is further assumed that their interactions are also described by such equations, no "clocks" existed immediately after the big bang, because according to relativity theory there is no passage of time for massless particles. Constructing a clock out of photons and gravitons alone is not possible, because they are conformally or effectively conformally invariant, respectively. Thus the early universe loses track of the scaling which determines the full spactime metric, while retaining its conformal geometry (rather than the slightly more restrictive metric geometry) – one might even claim that, near the big bang, no conception of the passage of time should be applied.

Similar conjectures about the distant future of the universe are reasonable too, but in a mathematically reverse sense: while a conformal factor becomes infinite at the big bang, it goes to zero at the future boundary, which is spacelike if there is a positive cosmological constant. So in the very remote future there are also no clocks in principle, and the universe "forgets" time. With conformal invariance both in the distant future and at the big bang, Penrose argues that the two situations are physically identical, so that the furthest future of one phase of the universe becomes the big bang of the next. Although temperature undergoes an enormous change at the crossover surface – approaching zero just before it and infinity immediately beyond it – the frequencies and thus energies of photons are completely rescaled. So without massive particles, there is no way of defining lengths or times, whence the only physically meaningful structure is the conformal structure, i.e., the causal structure. By compressing the conformal factor towards the far future, and expanding it towards the beginning, the geometry of the future conformal boundary can be joined seamlessly to the initial conformal boundary.

It is also crucial that the Weyl curvature should be zero on both the future boundary and the past boundary, so that the big bang is still well-defined in the model as the unique hypersurface on which the Weyl curvature vanishes. However, there are black hole spacetime regions of divergent Weyl curvature that are singularities in the conformal geometry, and these can be encountered as future endpoints of particles' timelike world lines. According to the Weyl curvature hypothesis, they indicate an asymmetry between past and future, just as a big crunch singularity does. This would be inconsistent with the conformal cyclic cosmology. Therefore the Weyl-divergent future-type singularities must have dissappeared by the time the future hypersurface is reached, in order not to violate the conformal smoothness. This seems possible indeed, because in the far future all black holes will eventually disappear due to the quantum effect of Hawking evaporation, which brings the Weyl curvature back to zero.

Penrose postulates that information is lost in this process and standard quantum theory is violated. Thereby entropy is also claimed to be effectively renormalized, decreasing significantly before crossing the hypersurface, which reproduces the enormous specialness of the big bang. This real information loss during or after black hole evaporation is in disagreement with other accounts [166], and it remains unclear and controversial whether the effective reduction of phase space volume violates the second law of thermodynamics, or even what this means.

The conformal cyclic cosmology is based on some other debatable assumptions. As in the pre-big bang scenario [53, 173] and in the cyclic universe scenario [140], there is no inflationary epoch – exponential expansion did not take place after the big bang but before it, and this generated the scale-invariant spectrum of density perturbations for the post-big bang universe. At the change-over from one aeon to the next, scalar fields might be produced, or simply remain; these would create dark matter after the big bang (which must eventually decay into massless particles in the remote future). And gravity is considered to be infinitely large at the big bang (which is why its degrees of freedom are zero, i.e., gravitational entropy is low), but gets smaller with time and eventually falls to zero at the final boundary. Furthermore, in the remote future there must be only electromagnetic and gravitational radiation, but no massive fermions and charged particles. This implies that protons will decay, neutrinos will become massless (or decay too), and electric charge is not exactly conserved (or electrons will also decay) – requiring an as yet unknown physical process. Only then does the conformal geometry become the relevant spacetime structure again. And quantum fluctuations, which are inevitable in a vacuum with positive energy density, must not spontaneously create anything from massive particles to black holes, which is expected on very large time-scales; otherwise they would prevent the universe from ever reaching an exact state of conformal invariance in the far future.

So conformal cyclic cosmology seems quite speculative. But it should be testable. For example, gravitational wave bursts arising from close encounters between black holes might leave their mark on the future boundary, influence its conformal geometry, and generate spatial variations in the matter density just after the next big bang. Such bursts of gravitational radiation would cause a superposition of circular patterns on the celestial cosmic microwave background sphere, similar to the appearance of ripples on a pond following a sustained period of rain (whether there are already hints for this is controversial, see [46, 47, 60–62, 64, 103, 145]).

In the conformal cyclic cosmology there is neither a multiversal "surface" of pseudo-beginning, in contrast to some quantum and string cosmological models of the emergent universe, nor is there a "branching" of universes in contrast to models such as eternal inflation, the recycling universe, or cosmological natural selection. The conformal cyclic cosmology represents a linear succession of universes (or one endlessly repeating kind of universe).

Penrose's model can also be interpreted as a kind of pseudo-beginning (and pseudo-ending) cosmology. It does not represent an absolute beginning because even when (macro)time disappears, space remains (at least if it is assumed that the big bang curvature singularities are overcome). Time, or macrotime, is not eternal, strictly speaking. The (thermodynamic) arrow of time vanishes, along with a supposed effective decrease of entropy – or phase space reduction –, if this really accompanies black hole evaporation. Therefore gaps in macrotime exist in the conformal cyclic cosmology, namely when there are no longer any massive particles around (in the remote future of each aeon) or when particles are effectively massless (in the very early phase of each aeon). But microtime continues. Causality remains, and there is not nothing – there are still lonely photons and spreading gravitational waves which may even influence the subsequent aeon.

Thus conformal cyclic cosmology can be categorized as lying between initial and eternal cosmologies – like some other models, but in a different way. There is a big bang beginning, in fact there are arbitrarily many, but it is not the beginning of everything, and it may even bear signs of the preceding aeon in the form of gravitational wave imprints. The succession of aeons seems to be eternal (if it is stable, which has not yet been shown), and may never vanish. But there are "chasms" in macrotime, i.e., recurrent disappearances of the thermodynamic arrow of time and other arrows – excepting the causal arrow, at least (con)formally. So there are "periods" where no clocks of any sort could be built and no passage of time could be observed in principle. In this scenario time ends temporarily, paradoxically speaking, at least if time is relational. (It may still be claimed, however, that there is a kind of fundamental, global time, pervading all the aeons, and at least conceptually this is somewhat presupposed by referring to an infinite succession of aeons.)

13 The End of Time

Whether time has a beginning or not and whether it ends at some point remains puzzling [170, 171].

If time is emergent, either created by fundamental physical entities or by our consciousness, it would ultimately be an illusion. And the end of time would have essentially arrived as soon as this illusion is debunked or explained (or as soon as consciousness comes to an end). Physicists like Julian Barbour [12] and Carlo Rovelli [125] advocate this idea and call it "the end of time". It is, strictly speaking, an epistemological or conceptual issue.

If time is fundamental, however, it remains an open question whether it is finite or not. This is a matter of physics and/or metaphysics.

An end of fundamental (and also emergent) time is equivalent to a classic global singularity – for example in the big crunch [16, 122] or in a sudden "freezing" of all dynamics, including cosmic expansion [23]. There would no longer be any classical time.

From the perspective of quantum mechanics, however, a wild superposition of states without any quasi-classical phenomena is more plausible [78]. This would be the end of the world as we know it, but not the end of everything.

Future singularities such as the global one in the big crunch or local ones in black holes might be avoided by means of quantum gravitational effects. Instead there would be a bounce in which time continues or in which it is reborn.

And there are other doomsday scenarios about how time could end. One is called big snap [144]. Here it is argued that – just like a rubber band, which cannot be stretched indefinitely because of its finite number of atoms – the granular nature of spacetime implies destructive events to come if space has a finite number of degrees of freedom and expands too much. Equally devastating would be a signature change [95, 96]. Here the time dimension turns to a fourth spatial dimension if the universe is a four-dimensional brane, moving through a higher-dimensional bulk, and approaches the velocity of light. This would cause a big freeze within the brane – though time would still continue within the bulk.

Another kind of dissolution of classical time could be classified as a pseudo-ending. It wouldn't have to be the end of everything, but it could lead to a quantum vacuum in which there would no longer exist a macrotime albeit there would still be a microtime – a structureless, reversible state of equilibrium. (Whether "existence" necessarily takes place "in time" or independently of time is a difficult terminological and philosophical question, but maybe not a scientific one.) So it seems possible that the arrows of time end – but not time itself. In other words: macrotime stops, but microtime goes on. This would be a pseudo-ending analogous to a pseudo-beginning of the universe. Here the end of time would mean the end of macrotime. This could happen if the future of space becomes an empty but eternally expanding de Sitter universe. (Only if it is fundamental would time remain, displaying itself in the expansion, although the latter cannot be measured.) Then there wouldn't be any more (irreversible) events, at least not locally, because there would be nothing left to change (neglecting virtual quantum processes).

Paradoxically, the end of (macro)time could be a temporary one if a vacuum is left. If random fluctuations above some threshold value arise, which in the long term are unavoidable in quantum systems, an arrow of time can develop again. This is somewhat similar to Boltzmann's [21] original fluctuation hypothesis, but it is based upon different physical conditions (quantum processes in a dynamical spacetime instead of mechanistically conceived atoms and radiation in an infinite static space; cf. [33]). In a quantum mechanical de Sitter universe, it is even possible for new big bangs to arise from the vacuum, which is not entirely empty due to quantum fluctuations (see [30, 37, 38, 113, 164]). Then new arrows of time emerge, and there is a time after time, a succession of "interrupted" macrotimes.

"There is a theory which states that if ever anybody discovers exactly what the Universe is for and why it is here, it will instantly disappear and be replaced by something even more bizarre and inexplicable. There is another theory which states that this has already happened", Douglas Adams once joked. Modern cosmology has come up with many strange models and possibilities. Issues of time are central here. Although it is too early to decide between the competing approaches, it seems quite probable that the universe is bizarre and perhaps even inexplicable and absurd to some extent [158]. But one thing seems to be certain already: time is not what it used to be.

Acknowledgements Though time might be an illusion, deadlines are not. For support, comments, and extraordinary patience I am very grateful to Martin Bojowald, Claus Kiefer, Angela Lahee, Stephen Lyle, Laura Mersini-Houghton, Andreas Müller, André Spiegel, Nela Varwig, Doug Washer, and Hans-Dieter Zeh. Thanks also to Karl Marx who helped me with Figure 5. Figures 1 and 2 were taken and modified from [169].

References

1. A. Aguirre, in *Beyond the Big Bang*, ed. by R. Vaas. Eternal Inflation (Springer, Heidelberg, 2012)
2. A. Aguirre, S. Gratton, Steady-state eternal inflation. Phys. Rev. D **65**, 083507 (2002), arXiv:astro-ph/0111191
3. M. Aiello, M. Castagnino, O. Lombardi, The arrow of time: from universe time-asymmetry to local irreversible processes. Found. Phys. **38**, 257–292 (2008), arXiv:gr-qc/0608099
4. A. Albrecht, in *Science and Ultimate Reality*, ed. by J.D. Barrow, P.C.W. Davies, C.L. Harper. Cosmic Inflation and the Arrow of Time (Cambridge University Press, Cambridge, 2004), 363–401, arXiv:astro-ph/0210527
5. J. Ambjorn, J. Jurkiewicz, R. Loll, Causal dynamical triangulations and the quest for quantum gravity (2010), arXiv:1004.0352
6. P.H. Aref'eva, S. Frampton, Matsuzaki, Multifluid Models for Cyclic Cosmology. Proc. Steklov Inst. Math. **265**, 59–62 (2009), arXiv:0802.1294
7. A. Ashtekar, Introduction to Loop Quantum Gravity (2012), arXiv:1201.4598
8. A. Ashtekar, M. Bojowald, in *Beyond the Big Bang*, ed. by R. Vaas. Loop Quantum Gravity and Cosmology (Springer, Heidelberg, 2012)
9. A. Ashtekar, M. Campiglia, A. Henderson, Casting loop quantum cosmology in the spin foam paradigm. Classical Quant. Grav. **27**, 135020 (2010), arXiv:1001.5147
10. A. Ashtekar, J. Lewandowski, Background independent quantum gravity: a status report (2004), arXiv:gr-qc/0404018
11. D. Atkatz, H.R. Pagels, Origin of the universe as a quantum tunneling event. Phys. Rev. D **25**, 2065–2072 (1982)
12. J. Barbour, *The End of Time* (Oxford University Press, Oxford, 2000)
13. J.D. Barrow, F.J. Tipler, *The Anthropic Cosmological Principle* (Oxford University Press, Oxford, 1986)
14. J.D. Barrow et al., On the stability of the Einstein static universe. Classical Quant. Grav. **20**, L155–L164 (2003), arXiv:gr-qc/0302094
15. C. Berger, L. Sehgal, CP violation and arrows of time evolution of a neutral K or B meson from an incoherent to a coherent state. Phys. Rev. D **76**, 036003 (2007), arXiv:0704.1232
16. G. Björnsson, E.H. Gudmundsson, Cosmological observations in a closed universe. Mon. Not. R. Astron. Soc. **274**, 793–807 (1995)
17. M. Bojowald, What happened before the big bang? Nature Phys. **3**, 523–525 (2007)
18. M. Bojowald, Loop quantum cosmology. Living Rev. Relativity **11**, 4 (2008), http://www.livingreviews.org/lrr-2008--4

19. M. Bojowald, Loop quantum gravity and cosmology: a dynamical introduction (2011), arXiv:1101.5592
20. M. Bojowald, A Momentous Arrow of Time (2012), this volume
21. L. Boltzmann, On certain questions of the theory of gases. Nature **51**, 413–415 (1985)
22. A. Borde, A.H. Guth, A. Vilenkin, Inflationary spacetimes are not past-complete. Phys. Rev. Lett. **90**, 151301 (2003), arXiv:gr-qc/0110012
23. M. Bouhmadi-Lopez, P.F. González-Díaz, P. Martín-Moruno, On the generalised Chaplygin gas: worse than a big rip or quieter than a sudden singularity? Int. J. Mod. Phys. D **17**, 2269–2290 (2008), arXiv:0707.2390
24. R. Bousso, Vacuum Structure and the Arrow of Time (2011), arXiv:1112.3341
25. R. Brout, F. Englert, E. Gunzig, The creation of the universe as a quantum phenomenon. Ann. Phys. **115**, 78–106 (1978)
26. R. Brout, F. Englert, P. Spindel, Cosmological origin of the grand-unification mass scale. Phys. Rev. Lett. **43**, 417–420 (1979)
27. M.G. Brown, K. Freese, W.H. Kinney, The Phantom Bounce: A New Oscillating Cosmology. JCAP **0803**, 002 (2008), arXiv:astro-ph/0405353
28. S. Brush, *The Kind of Motion We Call Heat* (North Holland, 1976)
29. S. Brush, *Statistical Physics and the Atomic Theory of Matter* (Princeton University Press, Princeton, 1983)
30. S.M. Carroll, J. Chen, Spontaneous inflation and the origin of the arrow of time (2004), arXiv:hep-th/0410270
31. S.M. Carroll, What if Time Really Exists? (2008), arXiv:0811.3772
32. S.M. Carroll, *From Eternity to Here* (Dutton, New York, 2010)
33. M.M. Cirkovic, The thermodynamical arrow of time: reinterpreting the Boltzmann–Schuetz argument. Found. Phys. **33**, 467–490 (2003), arXiv:astro-ph/0212511
34. B. Craps, O. Evnin, Adiabaticity and emergence of classical space–time in time-dependent matrix theories. J. High Energy Phys. **1101**, 130 (2011), arXiv:1011.0820
35. P.C.W. Davies, *The Physics of Time Asymmetry* (University of California Press, Berkeley, 1977)
36. A. De Simone, Boltzmann brains and the scale-factor cutoff measure of the multiverse. Phys. Rev. D **82**, 063520 (2010), arXiv:0808.3778
37. S. Dutta, T. Vachaspati, Islands in the lambda-sea: an alternative cosmological model. Phys. Rev. D **71**, 083507 (2005), arXiv:astro-ph/0501396
38. S. Dutta, T. Vachaspati, in *Beyond the Big Bang*, ed. by R. Vaas. Island Cosmology (Springer, Heidelberg, 2012)
39. L. Dyson, M. Kleban, L. Susskind, Disturbing implications of a cosmological constant. J. High Energy Phys. **0210**, 011 (2002), arXiv:hep-th/0208013
40. J. Earman, An attempt to add some direction to 'The problem of the direction of time'. Phil. Sci. **41**, 15–47 (1974)
41. A.S. Eddington, *The Nature of the Physical World* (Dutton, New York, 1927)
42. A. Einstein, Kosmologische Betrachtungen zur allgemeinen Relativitätstheorie. Königlich Preußische Akademie der Wissenschaften (Berlin), Sitzungsberichte (1917), pp. 142–152. Reprinted in *The Collected Papers of Albert Einstein*, vol. **6**, ed. by A.J. Knox, M.J. Klein, R. Schulmann (Princeton University Press, Princeton, 1996), pp. 541–552
43. G. Ellis, R. Maartens, The emergent universe: inflationary cosmology with no singularity. Classical Quant. Grav. **21**, 223–232 (2004), arXiv:gr-qc/0211082
44. G. Ellis, R. Maartens, in *Beyond the Big Bang*, ed. by R. Vaas. The Emergent Universe (Springer, Heidelberg, 2012)
45. G. Ellis, R. Maartens, J. Murugan, C. Tsagas, The emergent universe: an explicit construction. Classical Quant. Grav. **21**, 233–250 (2004), arXiv:gr-qc/0307112
46. H.K. Eriksen, I.K. Wehus, A search for concentric circles in the 7-year WMAP temperature sky maps. Astrophys. J. Lett. **733**, L29 (2011), arXiv:1012.1268
47. H.K. Eriksen, I.K. Wehus, Comment on "CCC-predicted low-variance circles in CMB sky and LCDM" (2011), arXiv:1105.1081

48. B. Falkenburg, *Kants Kosmologie* (Klostermann, Frankfurt am Main, 2000)
49. P.I. Fomin, in *Beyond the Big Bang*, ed. by R. Vaas. Gravitational Instability of the Vacuum and the Cosmological Problem (Springer, Heidelberg, 2012) (in Russian: 1973)
50. K. Freese, M.G. Brown, W.H. Kinney, The Phantom Bounce: A New Proposal for an Oscillating Cosmology (2012), this volume
51. B. Freivogel et al., Eternal Inflation, Bubble Collisions, and the Disintegration of the Persistence of Memory. JCAP **0908**, 036 (2009), arXiv:0901.0007
52. J. Garriga, A.H. Guth, A. Vilenkin, Eternal inflation, bubble collisions, and the persistence of memory. Phys. Rev. D **76**, 123512 (2007), arXiv:hep-th/0612242
53. M. Gasperini, G. Veneziano, The pre-big bang scenario in string cosmology. Phys. Reports **373**, 1–212 (2003), arXiv:hep-th/0207130
54. T. Gold, in *La Structure et l'Evolution de l'Univers. Proc. 11th International Solvay Congress*, ed. by R. Stoops. The Arrow of Time (Coudenberg, Brussels, 1958), pp. 81–95
55. T. Gold, The arrow of time. Am. J. Phys **30**, 403–410 (1962)
56. J.R. Gott III, Creation of open universes from de Sitter space. Nature **295**, 304–307 (1982)
57. J.R. Gott, L.X. Li, Can the universe create itself? Phys. Rev. D **58**, 023501 (1998), arXiv:astro-ph/9712344
58. J.R. Gott, L.X. Li, in *Beyond the Big Bang*, ed. by R. Vaas. The Self-Creating Universe (Springer, Heidelberg, 2012)
59. A. Grünbaum, in *Wege der Vernunft*, ed. by A. Bohnen, A. Musgrave. Die Schöpfung als Scheinproblem der physikalischen Kosmologie (Mohr, Tbingen, 1991), pp. 164–191
60. V.G. Gurzadyan, R. Penrose, Concentric circles in WMAP data may provide evidence of violent pre-big-bang activity (2010), arXiv:1011.3706
61. V.G. Gurzadyan, R. Penrose, More on the low variance circles in CMB sky (2010), arXiv:1012.1486
62. V.G. Gurzadyan, R. Penrose, CCC-predicted low-variance circles in CMB sky and LCDM (2011), arXiv:1104.5675
63. A.H. Guth, in *Cosmic Questions*, ed. by J.B. Miller. Eternal Inflation (New York Academy of Sciences, New York, 2001), pp. 66–82, arXiv:astro-ph/0101507
64. A. Hajian, Are There Echoes From The Pre-Big Bang Universe? A Search for Low Variance Circles in the CMB Sky (2010), arXiv:1012.1656
65. H. Haken, *Synergetics, an Introduction* (Springer, New York, 1983), 3rd rev. ed.
66. H. Haken, *Advanced Synergetics* (Springer, New York, 1993)
67. J.J. Halliwell, J. Perez-Mercader, W.H. Zurek (ed.), *Physical Origins of Time Asymmetry* (Cambridge University Press, Cambridge, 1994)
68. J.B. Hartle, The physics of 'Now'. Am. J. Phys. **73**, 101–109 (2005), arXiv:gr-qc/0403001
69. J.B. Hartle, S.W. Hawking, T. Hertog, The classical universes of the no-boundary quantum state. Phys. Rev. D **77**, 123537 (2008), arXiv:0803.1663
70. J. Hartle, T. Hertog, Arrows of time in the bouncing universes of the no-boundary quantum state (2011), arXiv:1104.1733
71. S.W. Hawking, Arrow of time in cosmology. Phys. Rev. D **32**, 2489–2495 (1985)
72. S.W. Hawking, R. Laflamme, G.W. Lyons, Origin of time asymmetry. Phys. Rev. D **47**, 5342–5356 (1993), arXiv:gr-qc/9301017
73. S.W. Hawking, *A Brief History of Time* (Bantam, New York, 1988)
74. S.W. Hawking, R. Penrose, *The Nature of Space and Time* (Princeton University Press, Princeton, 1996)
75. H. Heimsoeth, in *Studien zur Philosophiegeschichte. Zeitliche Weltunendlichkeit und das Problem des Anfangs* (Kölner Universitäts-Verlag, Köln, 1961), pp. 269–292 (original 1960)
76. D. Hilbert, Über das Unendliche. Math. Ann. **95**, 161–190 (1925)
77. M. Israelit, Primary matter creation in a Weyl–Dirac cosmological model. Found. Phys. **32**, 295–321 (2002)
78. A. Kamenshchik, C. Kiefer, B. Sandhoefer, Quantum cosmology with big-brake singularity. Phys. Rev. D **76**, 064032 (2007), arXiv:0705.1688
79. B. Kanitscheider, *Kosmologie* (Reclam, Stuttgart, 2002), 3rd rev. ed.

80. I. Kant, *Kritik der reinen Vernunft* (Suhrkamp, Frankfurt am Main 1990); engl.: *Critique of Pure Reason*, e.g., translated by N. Kemp Smith (first German publication 1781/1787)
81. A. Kashlinsky, F. Atrio-Barandela, H. Ebeling, Measuring the dark flow with public X-ray cluster data. Astrophys. J. **732**, 1 (2011), arXiv:1012.3214
82. C. Kiefer, H.D. Zeh, Arrow of time in a recollapsing quantum universe. Phys. Rev. D **51**, 4145–4153 (1995), arXiv:gr-qc/9402036
83. C. Kiefer, *Quantum Gravity* (Oxford University Press, Oxford, 2004)
84. C. Kiefer, Does time exist in quantum gravity? (2009), arXiv:0909.3767
85. C. Kiefer, Can the Arrow of Time Be Understood from Quantum Cosmology? (2012), this volume
86. H. Kragh, *Cosmology and Controversy* (Princeton University Press, Princeton, 1996)
87. J.-L. Lehners, Cosmic Bounces and Cyclic Universes (2011), arXiv:1106.0172
88. J.-L. Lehners, P.J. Steinhardt, N. Turok, The Return of the Phoenix Universe. Int. J. Mod. Phys. D **18**, 2231–2235 (2009), arXiv:0910.0834
89. A. Linde, Prospects of Inflation (2004), arXiv:hep-th/040
90. A. Linde, Sinks in the landscape, Boltzmann brains, and the cosmological constant problem. JCAP **0701**, 022 (2007), arXiv:hep-th/0611043
91. M.C. Mackey, *Time's Arrow: The Origins of Thermodynamic Behavior* (Springer, New York, 1992)
92. J.L. Mackie, *The Cement of the Universe* (Oxford University Press, Oxford, 1974)
93. W. Malzkorn, *Kants Kosmologie-Kritik* (de Gruyter Berlin, New York, 1999)
94. O.J.E. Maroney, Does a computer have an arrow of time? (2007), arXiv:0709.3131
95. M. Mars, J.M.M. Senovilla, R. Vera, Lorentzian and signature changing branes. Phys. Rev. D **76**, 044029 (2007), arXiv:0705.3380
96. M. Mars, J.M.M. Senovilla, R. Vera, Is the accelerated expansion evidence of a forthcoming change of signature on the brane? Phys. Rev. D **77**, 027501 (2008), arXiv:0710.0820
97. I. Masina, A. Notari, Detecting the cold spot as a void with the non-diagonal two-point function (2010), arXiv:1007.0204
98. T. Maudlin, Remarks on the passing of time. Proc. Arist. Soc. **CII**, 237–252 (2002)
99. B. McInnes, Arrow of time in string theory. Nucl. Phys. B **782**, 1–25 (2007), arXiv:hep-th/0611088
100. B. McInnes, in *Beyond the Big Bang*, ed. by R. Vaas. The arrow of time. (Springer, Heidelberg, 2012)
101. L. Mersini-Houghton, Is Eternal Inflation Eternal? (2011), arXiv:1106.3542
102. L. Mersini-Houghton, in *Beyond the Big Bang*, ed. by R. Vaas. Selection of initial conditions (Springer, Heidelberg, 2012)
103. A. Moss, D. Scott, J.P. Zibin, No evidence for anomalously low variance circles on the sky (2010), arXiv:1012.1305
104. M. Novello, S.E. Perez Bergliaffa, Bouncing Cosmologies. Phys. Reports **463**, 127–213 (2008), arXiv:0802.1634
105. R. Nozick, in *Philosophical Explanations*, Why is there something rather than nothing? (Clarendon, Oxford, 1981), pp. 668–679
106. D. Oriti, Spin foam models of quantum spacetime (2003), arXiv:gr-qc/0311066
107. D. Page, Will entropy decrease if the universe recollapses? Phys. Rev. D **32**, 2496–2499 (1985)
108. D.N. Page, Symmetric-bounce quantum state of the universe (2009), arXiv:0907.1893
109. H. Pagels, *Perfect Symmetry* (Bantam, New York, 1985)
110. C.M. Patton, J.A. Wheeler, in *Quantum gravity*, ed. by: C.J. Isham, R. Penrose, D.W. Sciama. Is physics legislated by cosmogony? (Clarendon, Oxford, 1975), pp. 538–605
111. R. Penrose, in *Quantum gravity*, ed. by C.J. Isham, R. Penrose, D.W. Sciama. Time-Asymmetry and Quantum Gravity, vol. **2** (Clarendon, Oxford, 1981), pp. 242–272
112. R. Penrose, *The Emperor's New Mind* (Oxford University Press, Oxford, 1989)
113. R. Penrose, *Cycles of Time* (Bodley Head, London, 2010)

114. A. Perez, Spin foam models for quantum gravity. Classical Quant. Grav. **20**, R43-R104 (2003), arXiv:gr-qc/0301113
115. V. Petkov, *Relativity and the Nature of Spacetime* (Springer, Heidelberg, 2005)
116. H. Price, *Time's Arrow and Archimedes' Point* (Oxford University Press, Oxford, 1996)
117. H. Price, B. Weslake, The time-asymmetry of causation (2008), http://philsci-archive.pitt.edu/4475/
118. I. Prigogine, *Vom Sein zum Werden*, (Piper, München, 1979)
119. E. Pöppel, *Grenzen des Bewußtseins*, (DVA, Stuttgart, 1985)
120. E. Pöppel, A hierarchical model of temporal perception. Trends Cogn. Sci. **1**, 56–61 (1997)
121. E. Rebhan, "Soft bang" instead of "big bang": model of an inflationary universe without singularities and with eternal physical past time. Astron. Astrophys. **353**, 1–9 (2000)
122. M.J. Rees, The collapse of the universe: an eschatological study. Observatory **89**, 193–198 (1969)
123. C. Rovelli, *Quantum Gravity* (Cambridge University Press, Cambridge, 2004)
124. C. Rovelli, Loop quantum gravity. Living Rev. Relativity **11**, 5 (2008), http://www.livingreviews.org/lrr-2008--5
125. C. Rovelli, "Forget Time" (2009), arXiv:0903.3832
126. C. Rovelli, Loop quantum gravity: the first twenty five years (2010), arXiv:1012.4707
127. L. Rudnick, S. Brown, L.R. Williams, Extragalactic radio sources and the WMAP cold spot. Astrophys. J. **671**, 40–44 (2007), arXiv:0704.0908
128. W.C. Salmon, *Causality and Explanation* (Oxford University Press, New York, 1998)
129. J. Schaffer, The Metaphysics of Causation (2008), http://plato.stanford.edu/archives/fall2008/entries/causation-metaphysics
130. J. Schmucker, *Das Weltproblem in Kants Kritik der reinen Vernunft* (Bouvier, Bonn, 1990)
131. L.S. Schulman, *Time's Arrows and Quantum Measurement* (Cambridge University Press, Cambridge, 1997)
132. L.S. Schulman, Opposite thermodynamic arrows of time. Phys. Rev. Lett. **83**, 5419–5422 (1999), arXiv:cond-mat/9911101
133. L.S. Schulman, A compromised arrow of time (2000), arXiv:cond-mat/0009139
134. L.S. Schulman, Resolution of causal paradoxes arising from opposing thermodynamic arrows of time. Phys. Lett. A **280**, 239–245 (2001), arXiv:cond-mat/0102014
135. Q. Smith, Kant and the Beginning of the World. The New Scholasticism **59**, 339–346 (1985)
136. Q. Smith, in *God and Time*, ed. by G. E, Ganssle, D. M. Woodruff. Time Was Created by a Timeless Point (Oxford University Press, Oxford, New York, 2002), pp. 95–128
137. L. Smolin, How far are we from the quantum theory of gravity? (2003), arXiv:hep-th/0303185
138. A.A. Starobinsky, A new type of isotropic cosmological models without singularity. Phys. Lett. B **91**, 99–102 (1980)
139. P.J. Steinhardt, The inflation debate: Is the theory at the heart of modern cosmology deeply flawed? Scientific American **304** (4), 18–25 (2011)
140. P. Steinhardt, in *Beyond the Big Bang*, ed. by R. Vaas. The cyclic universe (Springer, Heidelberg, 2012)
141. W.R. Stoeger, G.F.R. Ellis, U. Kirchner, Multiverses and cosmology: philosophical issues (2004), arXiv:astro-ph/0407329
142. L. Susskind, *The Cosmic Landscape* (Little Brown, New York, 2006)
143. L. Susskind, Fractal-Flows and Time's Arrow (2012), arXiv:1203.6440
144. M. Tegmark, How unitary cosmology generalizes thermodynamics and solves the inflationary entropy problem (2011), arXiv:1108.3080
145. P. Tod, Penrose's circles in the CMB and a test of inflation (2011), arXiv:1107.1421
146. E. Tryon, Is the universe a vacuum fluctuation? Nature **246**, 396–397 (1973)
147. R. Vaas, in *Lexikon der Neurowissenschaft*, vol. **4**. Zeit und Gehirn (Spektrum Akademischer, Heidelberg, 2001), pp. 154–167
148. R. Vaas, Wenn die Zeit rückwärts läuft. bild der wissenschaft. **12**, 46–55 (2002)
149. R. Vaas, Der Streit um die Willensfreiheit. Universitas **57**, 598–612, 807–819 (2002)
150. R. Vaas, Der kosmische Code. bild der wissenschaft. **12**, 40–46 (2003)

151. R. Vaas, Die Zeit vor dem Urknall. bild der wissenschaft. **4**, 60–67 (2003)
152. R. Vaas, Jenseits von Raum und Zeit. bild der wissenschaft. **4**, 50–56 (2003). Engl. translation: Beyond space and time, arXiv:physics/0401128
153. R. Vaas, in *Theologie und Kosmologie*, ed. by J. Hübner, I.O. Stamatescu, D. Weber. Ein Universum nach Maß? Kritische Überlegungen zum Anthropischen Prinzip in der Kosmologie, Naturphilosophie und Theologie (Mohr Siebeck, Tübingen, 2004), pp. 375–498
154. R. Vaas, Time before time (2004), arXiv:physics/0408111
155. R. Vaas, Das Duell: Strings gegen Schleifen. bild der wissenschaft. **4**, 44–49 (2004). Extended engl. translation: The duel: Strings versus loops; arXiv:physics/0403112
156. R. Vaas, Der Urknall aus fast nichts. bild der wissenschaft. **10**, 32–41 (2004)
157. R. Vaas, in *Wissenschaft, Religion und Recht*, ed. by E. Hilgendorf. Das Münchhausen-Trilemma in der Erkenntnistheorie, Kosmologie und Metaphysik (Logos, Berlin, 2006), pp. 441–474
158. R. Vaas, Aufrechtstehen im Nichts. Universitas **63**, 1118–1137, 1244–1259 (2008)
159. R. Vaas, Das Loch. bild der wissenschaft. **9**, 50–55 (2008)
160. R. Vaas, in *The Biological Evolution of Religious Mind and Behavior*, ed. by E. Voland, W. Schiefenhövel. Gods, gains, and genes. On the natural origin of religiosity by means of bio-cultural selection (Springer, Berlin, 2009), pp. 25–49
161. R. Vaas, Das wahnsinnige Universum. bild der wissenschaft. **3**, 58–61 (2009)
162. R. Vaas, Die Zeit vor der Zeit. Universitas **64**, 1124–1139 (2009)
163. R. Vaas, Multiverse scenarios in cosmology: classification, cause, challenge, controversy, and criticism. J. Cosmol. **4**, 666–676 (2010), arXiv:1001.0726
164. R. Vaas, Die ewige Wiederkehr der Zeit. bild der wissenschaft. **12**, 50–55 (2010)
165. R. Vaas, Der Dunkle Fluss. bild der wissenschaft. **5**, 50–55 (2010)
166. R. Vaas, *Tunnel durch Raum und Zeit*, 5th edn. (Kosmos, Stuttgart, 2012)
167. R. Vaas (ed.), *Beyond the Big Bang* (Springer, Heidelberg, 2012)
168. R. Vaas, in *Nietzsches Wissenschaftsphilosophie*, ed. by H. Heit, G. Abel, M. Brusotti. "Ewig rollt das Rad des Seins": Der "Ewige-Wiederkunfts-Gedanke" und seine Aktualität in der modernen physikalischen Kosmologie (de Gruyter, Berlin, 2012), pp. 371–399
169. R. Vaas, *Hawkings neues Universum*, 3rd edn. (Piper, München, 2011)
170. R. Vaas, in *Beyond the Big Bang*, ed. by R. Vaas. Eternal Existence (Springer, Heidelberg, 2012)
171. R. Vaas, Das Ende der Zukunft. bild der wissenschaft **1**, 46–53 (2012)
172. J.A. Vaccaro, T Violation and the Unidirectionality of Time. Found. Phys. **41**, 1569–1596 (2011), arXiv:0911.4528
173. G. Veneziano, M. Gasperini, in *Beyond the Big Bang*, ed. by R. Vaas. The Pre-Big Bang Scenario (Springer, Heidelberg, 2012)
174. A. Vilenkin, Creation of universes from nothing. Phys. Lett. B **117**, 25–28 (1982)
175. A. Vilenkin, *Many Worlds In One* (Hill and Wang, New York, 2006)
176. C.F. von Weizsäcker, Der zweite Hauptsatz und der Unterschied von Vergangenheit und Zukunft. Ann Phys **36**, 275–283 (1939)
177. R.M. Wald, The arrow of time and the initial conditions of the universe (2005), arXiv:gr-qc/0507094
178. V.S. Wike, *Kant's Antinomies of Reason* (University Press of America, Washington, 1982)
179. T.E. Wilkerson, *Kant's Critique of Pure Reason* (Clarendon Press, Oxford 1976)
180. S. Wolfram, *A New Kind of Science* (Wolfram Media, Champaign, 2002)
181. H.D. Zeh, *The Physical Basis of the Direction of Time* (Springer, Berlin, 2007), http://www.time-direction.de
182. H. D. Zeh, Open Questions Regarding the Arrow of Time (2012), this volume

Fundamental Loss of Quantum Coherence from Quantum Gravity

Rodolfo Gambini, Rafael A. Porto, and Jorge Pullin

Abstract We discuss the fundamental loss of unitarity that appears in quantum mechanics when one uses physically realistic devices to measure time and space. The effect is independent of any interaction with the environment and appears in addition to any usual environmental decoherence. We discuss the conceptual and potential experimental implications of this process of decoherence.

1 Introduction

In the usual formulation, quantum mechanics involves an idealization. The idealization is to assume that space and time can be measured with arbitrary accuracy. Devices to perform such a feat clearly do not exist in nature, since all measuring devices are subject to some level of quantum fluctuations. Therefore the equations of quantum mechanics, when cast in terms of the variables that are really measured in the laboratory, will differ from the traditional Schrödinger description. Although this is an idea that arises naturally in ordinary quantum mechanics, it is of paramount importance when one is discussing quantum gravity. Since general relativity is a

R. Gambini (✉)
Instituto de Física, Facultad de Ciencias, Universidad de la República, Iguá 4225, CP 11400 Montevideo, Uruguay
e-mail: rgambini@fisica.edu.uy

R.A. Porto
School of Natural Sciences, Institute for Advanced Study, Einstein Drive, Princeton, NJ 08540, USA
e-mail: rporto@ias.edu

J. Pullin
Department of Physics and Astronomy, Louisiana State University, Baton Rouge, LA 70803-4001, USA
e-mail: pullin@lsu.edu

L. Mersini-Houghton and R. Vaas (eds.), *The Arrows of Time*, Fundamental Theories of Physics 172, DOI 10.1007/978-3-642-23259-6_3,

generally covariant theory where one needs to describe the evolution in a relational way, one ends up describing how certain objects change when other objects, taken as clocks, change. At the quantum level a relational description will compare the outcomes of measurements of quantum objects. Quantum gravity is expected to be of importance in regimes (e.g. near the big bang or a black hole singularity) in which the assumption of the presence of a classical measuring devices is unrealistic. The question therefore arises: is the difference between the idealized version of quantum mechanics and the real one just of interest in situations when quantum gravity is predominant, or does it have implications in other settings? We will argue that indeed it does have wider implications. Some of them are relevant to conceptual questions (e.g. the problem of measurement in quantum mechanics or the black hole information paradox) and there might even be experimental implications.

A detailed discussion of these ideas can be found in previous papers [1–3], and in particular in the pedagogical review [4]. Here we present an abbreviated discussion.

The plan of this paper is as follows: in the next section we will discuss the form of the evolution in quantum mechanics when the time variable is measured by a real clock. In Sect. 3 we will consider a fundamental bound on how accurate can a real clock be and the implications it has for quantum mechanics in terms of real clocks and its consequences. Section 4 discusses the implications of the formalism.

2 Quantum Mechanics with Real Clocks

Given a physical system we want to study, we start by choosing a "clock". By this we mean a physical quantity (more precisely a set of quantities, like when one chooses a clock and a calendar to monitor periods of more than a day) that we will use to keep track of the passage of *time*. An example of such a variable could be the angular position of the hand of an analog watch. We denote it by T. We then identify some physical variables that we wish to study as a function of time. We shall call them generically O ("observables"). We then proceed to quantize the system by promoting all the observables and the clock variable to self-adjoint quantum operators acting on a Hilbert space. The latter is defined once a well defined inner product is chosen in the set of all physically allowed states. Usually it consists of squared integrable functions $\psi(q)$ with q the configuration variables.

Notice that the basis of all this is ordinary quantum mechanics, we are not modifying the underlying theory in any way. We assume that the system has an evolution in terms of an external parameter t, which is a classical variable, given by a Hamiltonian and with operators evolving with Heisenberg's equations (it is easier to present things in the Heisenberg picture, though it is not mandatory to use it for our construction). Then the standard rules of quantum mechanics and its probabilistic nature apply, but in terms of the variable t, which we will assume we cannot observe directly.

We will call the eigenvalues of the "clock" operator T and the eigenvalues of the "observables" O. We will assume that the clock and the measured system do not interact (if one considered an interaction it would produce additional effects to the one discussed). So the density matrix of the total system is a direct product of that of the system under study and the clock $\rho = \rho_{\text{sys}} \otimes \rho_{\text{cl}}$, and the system evolves through a unitary evolution operator that is of the tensor product form $U = U_{\text{sys}} \otimes U_{\text{cl}}$. The quantum states are described by a density matrices at a time t. Since the latter is unobservable, we would like to shift to a description where we have density matrices as functions of the observable time T. We define the probability that the resulting measurement of the clock variable T correspond to the value t,

$$\mathcal{P}_t(T) \equiv \frac{\text{Tr}\left(P_T(0)U_{\text{cl}}(t)\rho_{\text{cl}}U_{\text{cl}}(t)^\dagger\right)}{\int_{-\infty}^{\infty} dt\, \text{Tr}\left(P_T(t)\rho_{\text{cl}}\right)}, \tag{1}$$

where $P_T(0)$ is the projector on the eigenspace with eigenvalue T evaluated at $t = 0$. We note that $\int_{-\infty}^{\infty} dt \mathcal{P}_t(T) = 1$. We now define the evolution of the density matrix,

$$\rho(T) \equiv \int_{-\infty}^{\infty} U_{\text{sys}}(t)\rho_{\text{sys}}U_{\text{sys}}(t)^\dagger \mathcal{P}_t(T) \tag{2}$$

where we dropped the "sys" subscript in the left hand side since it is obvious we are ultimately interested in the density matrix of the system under study, not that of the clock.

We have therefore ended with an "effective" density matrix in the Schrödinger picture given by $\rho(T)$. It is possible to reconstruct entirely in a relational picture the probabilities using this effective density matrix, for details we refer the reader to the lengthier discussion in [4]. By its very definition, it is immediate to see that in the resulting evolution unitarity is lost, since one ends up with a density matrix that is a superposition of density matrices associated with different t's and that each evolve unitarily according to ordinary quantum mechanics.

Now that we have identified what will play the role of a density matrix in terms of a "real clock" evolution, we would like to see what happens if we assume the "real clock" is behaving semi-classically. To do this we assume that $\mathcal{P}_t(T) = f(T - T_{\text{max}}(t))$, where f is a function that decays very rapidly for values of T far from the maximum of the probability distribution T_{max}. We refer the reader to [4] for a derivation, but the resulting evolution equation for the probabilities is,

$$\frac{\partial\rho(T)}{\partial T} = i[\rho(T), H] + \sigma(T)[H, [H, \rho(T)]]. \tag{3}$$

and the extra term is dominated by the rate of change $\sigma(T)$ of the width of the distribution $f(t - T_{\text{max}})(t)$.

An equation of a form more general than this has been considered in the context of decoherence due to environmental effects, it is called the Lindblad equation. Our particular form of the equation is such that conserved quantities are automatically preserved by the modified evolution. Other mechanisms of decoherence coming from a different set of effects of quantum gravity have been criticized in the past because they fail to conserve energy [5]. Our approach does not suffer from those problems. It should be noted that Milburn arrived at a similar equation as ours from different assumptions [6]. Egusquiza, Garay and Raya derived a similar expression from considering imperfections in the clock due to thermal fluctuations [7]. It is to be noted that all such effects will occur in addition to the ones we discuss here. Corrections to the Schrödinger equation from quantum gravity have also been considered in the context of WKB analyses [8].

In a real experiment, there will be decoherence in the system under study due to interactions with the environment, that will be superposed on the effect we discuss. Such interactions might be reduced by cleverly setting up the experiment. The decoherence we are discussing here however, is completely determined by the quality of the clock used. It is clear that if one does experiments in quantum mechanics with poor clocks, pure states will evolve into mixed states very rapidly. The effect we are discussing can therefore be made arbitrarily large simply by choosing a lousy clock. Similar effects have actually been observed experimentally in the Rabi oscillations describing the exchange of excitations between atoms and field [9].

3 Fundamental Limits to Realistic Clocks

We have established that when we study quantum mechanics with a physical clock (a clock that includes quantum fluctuations), unitarity is lost, conserved quantities are still preserved, and pure states evolve into mixed states. The effects are more pronounced the worse the clock is. Which raises the question: is there a fundamental limitation to how good a clock can be? This question was first addressed by Salecker and Wigner [10]. Their reasoning went as follows: suppose we want to build the best clock we can. We start by insulating it from interactions with the environment. An elementary clock can be built by considering light bouncing between two mirrors. The clock "ticks" every time light strikes one of the mirrors. Even completely isolating the clock from any environmental effects, it develops errors. The reason for them is that by the time the photon travels between the mirrors, the wavefunctions of the mirrors spread. Therefore the time of arrival of the photon develops an uncertainty. Salecker and Wigner calculated the uncertainty to be $\delta t \sim \sqrt{t/M}$ where M is the mass of the mirrors and t is the time to be measured (we are using units where $\hbar = c = 1$ and therefore mass is measured in 1/s). The longer the time measured the larger the error. The larger the mass of the clock, the smaller the error.

So this tells us that one can build an arbitrarily accurate clock just by increasing its mass. However, Ng and Van Dam [11] pointed out that there is a limit to this. Basically, if one piles up enough mass in a concentrated region of space one ends up with a black hole. Some readers may ponder why do we need to consider a concentrated region of space. The reason is that if we allow the clock to be more massive by making it bigger, it also loses accuracy due to finite travel time of interactions in matter (see the discussion in [12] in response to [13]).

A black hole can be thought of as a clock since it is an oscillator. In fact it is the "fastest" oscillator one can have, and therefore the best clock for a given size. It has normal modes of vibration that have frequencies that are of the order of the light travel time across the Schwarzschild radius of the black hole. The more mass in the black hole, the lower the frequency, and therefore the worse its performance as a clock. This therefore creates a tension with the argument of Salecker and Wigner, which required more mass to increase the accuracy. This indicates that there actually is a "optimal spot" in terms of the mass that minimizes the error. Taking this into account one finds that the best accuracy one can get in a clock is given by $\delta T \sim T_{\text{Planck}}^{2/3} T^{1/3}$ where $T_{\text{Planck}} = 10^{-44}s$ is Planck's time and T is the time interval to be measured. This is an interesting result. On the one hand it is small enough for ordinary times that it will not interfere with most known physics. On the other hand is barely big enough that one might contemplate experimentally testing it, perhaps in future years.

With this absolute limit on the accuracy of a clock we can quickly work out an expression for the $\sigma(T)$ that we discussed in the previous section [3, 14]. With this estimate of the absolute best accuracy of a clock, we can work out again the evolution of the density matrix for a physical system in the energy eigenbasis. One gets

$$\rho(T)_{nm} = \rho_{nm}(0)e^{-i\omega_{nm}T} e^{-\omega_{nm}^2 T_{\text{Planck}}^{4/3} T^{2/3}}. \tag{4}$$

So we conclude that *any* physical system that we study in the lab will suffer loss of quantum coherence at least at the rate given by the formula above. This is a fundamental, inescapable limit. A pure state inevitably will become a mixed state due to the impossibility of having a perfect classical clock in nature to keep track of things.

4 Possible Experimental Implications

One can ask what are the prospects for detecting the fundamental decoherence we propose. At first one would expect them to be dim. It is, like all quantum gravitational effects, an "order Planck" effect. But it should be noted that the factor accompanying the Planck time can be rather large. For instance, if one would like to observe the effect in the lab one would require that the decoherence manifest itself in times of the order of magnitude of hours, perhaps days at best. That requires energy differences of the order of $10^{10}eV$ in the Bohr frequencies of the system. Such energy differences can only be achieved in "Schrödinger cat" type experiments, but are not outrageously beyond our present capabilities. Among the best candidates today are Bose–Einstein condensates, which can have 10^6 atoms in coherent states. However, it is clear that the technology is still not there to actually detect these effects, although it could be possible in forthcoming years.

A point that could be raised is that atomic clocks currently have an accuracy that is less than a decade of orders of magnitude worse than the absolute limit we derived in the previous section. Could not improvements in atomic clock technology actually get better than our supposed absolute limit? This appears unlikely. When one studies in detail the most recent proposals to improve atomic clocks, they require the use of entangled states [15] that have to remain coherent. Our effect would actually prevent the improvement of atomic clocks beyond the absolute limit!

Another point to be emphasized is that our approach has been quite naive in the sense that we have kept the discussion entirely in terms of non-relativistic quantum mechanics with a unique time across space. It is clear that in addition to the decoherence effect we discuss here, there will also be decoherence spatially due to the fact that one cannot have clocks perfectly synchronized across space and also that there will be fundamental uncertainties in the determination of spatial positions. This is discussed in some detail in our paper [16].

5 Conceptual Implications

The fact that pure states evolve naturally into mixed states has conceptual implications in at least three interesting areas of physics. We will discuss them separately.

5.1 The Black Hole Information Paradox

The black hole information paradox appeared when Hawking [17] noted that when quantum effects are taken into account, black holes emit radiation like a black body with a temperature $T_{\rm BH} = \hbar/(8\pi kGM)$ where M is the black hole mass, k is Boltzmann's constant and G is Newton's constant. As the black hole radiates, it loses mass, and therefore its temperature increases. This process continues until the black hole eventually evaporates completely and the only thing left is outgoing purely thermal radiation. Now, suppose one had started with a pure quantum state of enough mass that it collapses into a black hole. After the evaporation process, one is left with a mixed state (the outgoing purely thermal radiation). In ordinary quantum mechanics this presents a problem, since pure states cannot evolve into mixed states. (For further discussion and references on the paradox see [18]).

On the other hand, we have argued that due to the lack of perfectly classical clocks, quantum mechanics really implies that pure states do evolve into mixed states. The question is: could the effect be fast enough to render the black hole information paradox effectively unobservable? Our effect is small. But it is also true that black holes usually take a very long time to evaporate. Of course, a full calculation of the evaporation of a black hole would require a detailed modeling including quantum effects of gravity that no one is in a position of carrying out yet. We have done a very naive estimate [3, 14] of how our effect would take place in the case of an evaporating black hole. To this aim we have assumed the black hole is a system with energy levels (this is a common assumption in many quantum gravity scenarios), and that most of the Hawking radiation is coming from a transition between two dominant energy levels separated by a characteristic frequency dependent on the temperature. A detailed calculation based on this naive model [3] for the evolution of the density matrix shows that,

$$|\rho_{12}(T_{\max})| \sim |\rho_{12}(0)| \left(\frac{M_P}{M_{\rm BH}}\right)^{\frac{2}{3}}. \quad (5)$$

For astrophysical sized black holes, where $M_{\rm BH}$ is of the order of the mass of the Sun, this indicates that the off diagonal elements are suppressed by the time of evaporation by 10^{-28}, rendering the information puzzle effectively unobservable. What happens for smaller black holes? The effect is smaller. So can one claim that there still is an observable information puzzle for smaller black holes? This is debatable. After all, we do expect decoherence from other environmental effects to be considerably larger than the one we are considering here. If one makes the holes too small, then none of these calculations apply, and in fact the traditional Hawking evaporation is not an adequate description, since one has to take into account full quantum gravity effects. A better calculation than the one we did could probably be attempted, since both in string theory and loop quantum gravity there is some understanding of the energy levels of a black hole, even though the evaporation process is not well understood. Using such levels one could get a better estimate of how much coherence is lost. An interesting observation is that in certain recent matrix models of black hole evaporation in AdS black holes, correlators exhibit "revivals" in future times that our effect would clearly suppress in the long run [19].

5.2 The Measurement Problem in Quantum Mechanics

A potential conceptual application of the fundamental decoherence that we discussed is in connection with the measurement problem in quantum mechanics. The latter is related to the fact that in ordinary quantum mechanics the measurement apparatus is assumed to be always in an eigenstate after a measurement has been performed. The usual explanation [20] for this is that there exists interaction with the environment. This selects a preferred basis, i.e., a particular set of quasi-classical states that commute, at least approximately, with the Hamiltonian governing the system-environment interaction. Since the form of the interaction Hamiltonians usually depends on familiar "classical" quantities, the preferred states will typically also correspond to the small set of "classical" properties. Decoherence then quickly damps superpositions between the localized preferred states when only the system is considered. This is taken as an explanation of the appearance to a local observer of a "classical" world of determinate, "objective" (robust) properties.

The main problem with such a point of view is how is one to interpret the local suppression of interference in spite of the fact that the total state describing the system-environment combination retains full coherence. One may raise the question whether retention of the full coherence could ever lead to empirical conflicts with the ascription of definite values to macroscopic systems. The usual point of view is that it would be very difficult to reconstruct the off diagonal elements of the density matrix in practical circumstances. However, at least as a matter of principle, one could indeed reconstruct such terms (the evolution of the whole system remains unitary [21]) by "waiting long enough".

Our mechanism of fundamental decoherence could contribute to the understanding of this issue, since it implies that coherence is irreversibly lost and therefore one cannot reconstruct the off diagonal elements. Some people claim that we have just changed the environment by the clock as responsible for the loss of coherence and therefore the original criticism applies. But in the case of the clock, the minimum "size" of it in terms of its degrees of freedom if one wishes to view it as "a particular form of an environment" is determined by the length of the experiment and guarantees that in that length one will not be able to reconstruct the off diagonal elements. There is not the luxury of "waiting long enough" in this setting. For further discussion see our paper [22].

If one can, via this construction, end up with a quantum mechanics without a reduction postulate, this opens intriguing conceptual questions: such a theory would have the same predictions as ordinary quantum mechanics with a reduction postulate, yet it is completely different conceptually. This would imply that the nature of reality becomes undecidable. An initial discussion of this possibility is in a recent paper of ours [23].

5.3 *Quantum Computing*

In quantum computing, when one performs operations one is evolving quantum states. If one wishes the computers to perform faster, one needs to expend extra energy to evolve the quantum states. Based on this premise, Lloyd [24] presented a fundamental limitation to how fast quantum computers can be. Using the Margolus–Levitin [25] theorem he notes that in order to perform a computation in a time ΔT one needs to expend at least an energy $E \geq \pi\hbar/(2\Delta T)$. As a consequence, a system with an average energy E can perform a maximum of $n = 2E/(\pi\hbar)$ operations per second. For an "ultimate laptop" (a computer of a volume of one liter and one kilogram of weight) the limit turns out to be 10^{51} operations per second.

Such results assume the evolution is unitary. When it is not, as we have argued in this paper, erroneous computations are carried out. Since the rate of decoherence we discussed increases with increased energy differences, the rate of erroneous computations increases the faster one wishes to make the computer.

Can't one error correct? After all, one expects quantum computers to have errors due to decoherence from environmental factors. One can indeed error-correct.

But there are limitations to how fast this can be done. At its most basic level error correction is achieved by duplicating calculations and comparing results. This requires spatial communication, which is limited by the speed of light. Our point is that one cannot simply error correct one's way out of the fundamental decoherence effects.

We have to distinguish a bit between serial and parallel computing. In serial computing one achieves speed by increasing the energy in each qubit. This enhances our decoherence effect and significantly affects the performance. In a parallel machine one increases the speed by operating simultaneously on many qubits with lower energies per qubit therefore lowering the importance of the effect we introduced. For a machine with L qubits and a number of simultaneous operations d_P one gets,

$$n \le \left(\frac{1}{t_{\mathrm{P}}}\right)^{4/7} \left(\frac{cL}{R}\right)^{3/7} d_p^{4/7} \sim 10^{47}\mathrm{op/s}, \tag{6}$$

where the last estimate was obtained by taking the values of parameters for the "ultimate laptop" (for more details see [26]).

This is actually four orders of magnitude stronger than the bound that Lloyd found. If one had chosen a serial machine, the bound would have been tighter, 10^{42} operations per second.

We therefore see that although the effect we introduced is far from being achievable in quantum computers built in the next few years, it can limit the ultimate computing power of quantum computer. This is quite remarkable, given that it is a limit obtained involving gravity. Few people could have foreseen that gravity would play any role in quantum computation.

6 Discussion

We have argued that the use of realistic clocks and rods to measure space and time in quantum mechanics implies that pure states evolve into mixed states. Another way of putting this is that we are allowing quantum fluctuations in our clock. Similar ideas have been considered by Bonifacio, with a different formulation [27]. In quantum gravity and quantum cosmology it is natural to consider the clock to be part of the system under study. This is what motivated our interest in these issues, but it is clear that the core of the phenomenon can be described without references to quantum gravity, and that is what we have attempted to do in this presentation.

Even in the absence with current technology of a possibility of directly detecting these effects, they can have important conceptual implications, as we have illustrated with the black hole information puzzle, quantum computing and the problem of measurement in quantum mechanics.

Acknowledgements This work was supported in part by grants NSF-PHY0650715, and by funds of the Horace C. Hearne Jr. Institute for Theoretical Physics, PEDECIBA (Uruguay), FQXi, the University of California and CCT-LSU.

References

1. R. Gambini, R. Porto, J. Pullin, Class. Quant. Grav. **21**, L51 (2004) [arXiv:gr-qc/0305098]
2. R. Gambini, R. Porto, J. Pullin, New J. Phys. **6**, 45 (2004) [arXiv:gr-qc/0402118]
3. R. Gambini, R. Porto, J. Pullin, Braz. J. Phys. **35**, 266 (2005) [arXiv:gr-qc/0501027]
4. R. Gambini, R. Porto, J. Pullin, Gen. Rel. Grav. **39**, 1143 (2007) [arXiv:gr-qc/0603090]
5. See for instance, J. Ellis, J. Hagelin, D.V. Nanopoulos, M. Srednicki, Nucl. Phys. **B241**, 381 (1984); T. Banks, M.E. Peskin, L. Susskind, Nucl. Phys. **B244**, 125 (1984)
6. G.J. Milburn, Phys. Rev. **A44**, 5401 (1991)
7. I. Egusquiza, L. Garay, J. Raya, Phys. Rev. **A59**, 3236 (1999) [arXiv:quant-ph/9811009]
8. C. Kiefer, T. Singh, Phys. Rev. **D44**, 1061 (1991)
9. D.M. Meekhof, C. Monroe, B.E. King, W.M. Itano, D.J. Wineland, Phys. Rev. Lett. **76**, 1796 (1996); M. Brune, F. Schmidt-Kaler, A. Maali, J. Dreyer, E. Hagley, J.M. Raimond, S. Haroche, Phys. Rev. Lett. **76**, 1800 (1996); R. Bonifacio, S. Olivares, P. Tombesi, D. Vitali, Phys. Rev. **A61**, 053802 (2000)
10. E. Wigner, Rev. Mod. Phys. **29**, 255 (1957)
11. Y.J. Ng, H. van Dam, Ann. N. Y. Acad. Sci. **755**, 579 (1995) [arXiv:hep-th/9406110]; Mod. Phys. Lett. A **9**, 335 (1994)
12. Y.J. Ng, H. van Dam, Class. Quant. Grav. **20**, 393 (2003) [arXiv:gr-qc/0209021]
13. J.C. Baez, S.J. Olson, Class. Quant. Grav. **19**, L121 (2002) [arXiv:gr-qc/0201030]
14. R. Gambini, R.A. Porto, J. Pullin, Phys. Rev. Lett. **93**, 240401 (2004) [arXiv:hep-th/0406260]
15. See for instance, A. Andre, A. Sorensen, M. Lukin, Phys. Rev. Lett **92**, 230801 (2004) [arXiv:quant-ph/0401130]
16. R. Gambini, R.A. Porto, J. Pullin, Int. J. Mod. Phys. D **15**, 2181 (2006) [arXiv:gr-qc/0611148]
17. S. Hawking, Commun. Math. Phys. **43**, 199 (1975)
18. See for instance, S. Giddings, L. Thorlacius, in *Particle and Nuclear Astrophysics and Cosmology in the Next Millennium*, ed. by E. Kolb (World Scientific, Singapore, 1996) [arXiv:astro-ph/9412046]; for more recent references see S.B. Giddings, M. Lippert, [arXiv:hep-th/0402073]; D. Gottesman, and J. Preskill, J. High Energ. Phys. **0403**, 026 (2004) [arXiv:hep-th/0311269]
19. N. Iizuka, J. Polchinski, JHEP **0810**, 028 (2008) [arXiv:0801.3657 [hep-th]]
20. M. Schlosshauer, Rev. Mod. Phys. **76**, 1267 (2004) [arXiv:quant-ph/0312059]
21. R. Omnès, *The interpretation of quantum mechanics*, Princeton Series in Physics (Princeton, NJ, 1994)
22. R. Gambini and J. Pullin, Found. Phys. **37**, 1074 (2007)
23. R. Gambini, J. Pullin, in *Minkowski spacetime, a hundred years later*, V. Petkov (ed.), Springer, New York (2010) [arXiv:0801.2564 [gr-qc]]
24. S. Lloyd, Nature. **406**, 1047 (2000)
25. N. Margolus, L. Levitin, Physica. **D120**, 188 (1998)
26. R. Gambini, R. Porto, J. Pullin, in *Gravity, Astrophysics and Strings at the Black Sea*, ed. by P. Fiziev, M. Todorov (St. Kliment Ohridski Press, Sofia, 2006) [arXiv:quant-ph/0507262]
27. R. Bonifacio, Nuo. Cim. **D114**, 473 (1999)

The Clock Ambiguity: Implications and New Developments

Andreas Albrecht and Alberto Iglesias

Abstract We consider the ambiguity associated with the choice of clock in time reparameterization invariant theories. This arbitrariness undermines the goal of prescribing a fixed set of physical laws, since a change of time variable can completely alter the predictions of the theory. We review the main features of the clock ambiguity and our earlier work on its implications for the emergence of physical laws in a statistical manner. We also present a number of new results: We show that (contrary to suggestions in our earlier work) time independent Hamiltonians may quite generally be assumed for laws of physics that emerge in this picture. We also further explore the degree to which the observed Universe can be well approximated by a random Hamiltonian. We discuss the possibility of predicting the dimensionality of space, and also relate the second derivative of the density of states to the heat capacity of the Universe. This new work adds to the viability of our proposal that strong predictions for physical laws may emerge based on statistical arguments despite the clock ambiguity, but many open questions remain.

1 Introduction

Every theory that is invariant under time reparameterization presents a problem the moment we attempt quantization. Quantization gives a preferential role to time (in the definition of canonical variables) that cannot be fulfilled in a theory that is unaltered by its reparameterization. A prominent example of such a theory is given by General Relativity and in this context there have been extensive discussions of the problem (see, for example, [1] for an early treatment or [2] for an comprehensive review). An approach often used in cosmology is to work in "superspace" finding

A. Albrecht (✉) · A. Iglesias
Department of Physics, University of California at Davis, One Shields Avenue, Davis, CA 95616, USA
e-mail: ajalbrecht@ucdavis.edu

L. Mersini-Houghton and R. Vaas (eds.), *The Arrows of Time*, Fundamental Theories of Physics 172, DOI 10.1007/978-3-642-23259-6_4,

time as an "internal" variable after quantization. The invariance is imposed on the quantum states of the superspace $|\psi\rangle_S$ as a physical condition involving the Hamiltonian constraint,

$$\mathscr{H}|\psi\rangle_S = 0. \tag{1}$$

In [3, 4], we argued that such an approach carries an intrinsic arbitrariness in the choice of "clock" subspace that leads in turn to an arbitrariness in the predictions of the theory; the clock ambiguity. We showed that its implications are so severe that we may need to see the laws of physics as we know them as an approximate emergent phenomenon.

By taking the clock ambiguity seriously, we look for the emergence of physical properties derived from a Hamiltonian evolution chosen randomly, corresponding to an absolute ambiguity in the choice of clock. In [4] we singled out quasi-separability as a crucial feature of physical laws needed to sustain observers, and argued that quasi-separability is optimally achieved through locality (and thus through local field theory). In that context, we find our result from [4] that any sufficiently large random Hamiltonian can be interpreted (to a sufficiently good approximation) as a local field theory encouraging: It suggests that combining the randomness suggested by the clock ambiguity with the need for quasi-separability could yield local field theory as a *prediction*.

In this work, Sect. 2 reviews the clock ambiguity and sketches the basic approach we advocated in [4] to seek predictive power based on a statistical analysis. Section 3 gives a new result that shows that one can quite generally take the physical laws that emerge in our analysis to have a time *independent* Hamiltonian (this result is in contrast to assumptions we made in our earlier work). Section 4 reviews our analysis from [4] showing that any sufficiently large random Hamiltonian can be interpreted, to a good approximation, as a local field theory. In Sect. 4.2 we extend that work to discuss the possibility of predicting the dimensionality of space, and apply our analysis to a non-standard distribution of random Hamiltonians in Sect. 4.3, with interesting implications for higher orders in our Taylor series comparison of random Hamiltonians with field theories. After reviewing our thinking about gravity in this picture in Sect. 5, we extend our treatment to gravitating systems in Sect. 6 by relating the derivatives of the density of states to appropriate thermodynamic quantities which can be estimated for gravitating systems. The result of this extension, while very crude, is encouraging. We present our conclusions in Sect. 7

2 Summary of the Clock Ambiguity

The clock ambiguity arises from the treatment of time as "internal" in time reparameterization invariant theories. "Internal time" means that a subsystem of the universe is identified as the time parameter or "clock" and time evolution is revealed by examining correlations between the rest of the universe and the clock subsystem. In quantum theories this picture is typically expressed in "superspace", of which the clock system is a subspace.

In previous work [3, 4] we pointed out that regardless of how careful one is to describe a universe as obeying specific physical laws, the same state in the same superspace can equally well describe a completely different physical world with completely different time evolution. One only has to identity a different clock subsystem to find this new description. This is the clock ambiguity. We have shown that the clock ambiguity is absolute, in the sense that all possible systems experiencing all possible time evolution can be extracted from the same superspace state by a suitable choice of clock.

We refer the reader to this earlier work for the details [3, 4]. Here we quote the main result. We assume a discrete formalism which allows us to write the state in superspace as

$$|\psi\rangle_S = \sum_{ij} \alpha_{ij} \, |t_i\rangle_C \, |j\rangle_R \equiv \sum_i |t_i\rangle_C \, |\phi_i\rangle_R. \tag{2}$$

Here the subscripts S, C and R relate to the decomposition of superspace S according to $S = C \otimes R$, and refer to the full superspace, the clock subspace and the "rest" of the superspace respectively. The bases $|t_i\rangle_C$ and $|j\rangle_R$ span the clock and "rest" subspaces. The second equality defines (by summing over j) $|\phi_i\rangle_R$, giving the wavefunctions of the "rest" subspace at times t_i.

One can see that all the information about the state in the R subspace and its time evolution is contained within the expansion coefficients α_{ij}. In [3, 4] we show that arbitrary values α'_{ij} can result from expressing the same superspace state $|\psi\rangle_S$ according to suitable choices of the decomposition $S = C' \otimes R'$, or in other words, by making a suitable choice of clock. Thus any state evolving according to any Hamiltonian can be found, merely by choosing a new clock in the superspace.

One possible conclusion from the clock ambiguity is that the formalism that leads to this result must be wrong in some way (that in itself would have interesting implications). Otherwise, if we conclude that our fundamental theories really must have the clock ambiguity, the success of physics so far implies that it must be possible to come up with sharp predictions of specific physical laws, presumably based on some kind of statistical arguments, given that all possible physical laws are represented in the formalism. Note that, interestingly, Chris Wetterich has considered very similar issues using the functional integral formalism [5, 6].

In [4] we explored how one might go about formulating such a statistical analysis, and gave special emphasis to the quasi-separability of physical laws which seems so crucial for our ability to survive and thrive as tiny observers. We noted that locality (as realized in the local field theories that describe the elementary particles and forces) is the ultimate origin of the quasi-separability we experience in our physical world. We also noted that in some sense local field theories give a maximal expression of quasi-locality. Thus we feel our result from [4], that any random Hamiltonian can yield a sufficiently good approximation to a local field theory is quite interesting. It suggests that the requirement of quasi-separability may universally lead to local field theories as one searches for emergent physical laws in theories with the clock ambiguity. We review and extend that result in Sect. 4.

3 The Time Independence of H

A randomly chosen clock leads to a randomly chosen set of α_{ij}'s. Random α_{ij}'s describe a randomly chosen state evolving under a random Hamiltonian. The lack of any a priori reason to expect correlations between the α_{ij}'s with different i values suggests that in general the random Hamiltonian will be different for each time step (labeled by i). We discuss this issue in Sect. III-B of [4].

However, in our earlier work we overlooked a rather simple point (kindly brought to our attention by Glenn Starkman, 2007, private communication). The point is that α_{ij}'s do not contain nearly enough information to specify a full Hamiltonian at each time. We can use this fact to add a requirement that the Hamiltonian is time independent without any loss of generality, assuming one does not take too many time steps. We show below that this constraint is very easy to meet.

A time step can be written as

$$|\psi(t_{i+1})\rangle_R = [\mathbf{1} - i\hbar(\Delta t_i)\mathbf{H}(t_i)]\,|\psi(t_i)\rangle_R\,. \tag{3}$$

By taking the inner product of this equation with ${}_R\langle\psi(t_{i+1})|$ one finds

$$1 = {}_R\langle\psi\,(t_{i+1})\,|\mathbf{1} - i\hbar\,(\Delta t_i)\,\mathbf{H}\,(t_i)\,|\psi\,(t_i)\rangle_R. \tag{4}$$

The inner product with ${}_R\langle\psi^{\perp}(t_{i+1})|$ gives

$$0 = {}_R\langle\psi^{\perp}\,(t_{i+1})\,|\mathbf{1} - i\hbar\,(\Delta t_i)\,\mathbf{H}\,(t_i)|\,\psi\,(t_i)\rangle_R \tag{5}$$

where ${}_R\langle\psi^{\perp}(t_{i+1})|$ could be any one of $N-1$ states orthogonal to ${}_R\langle\psi(t_{i+1})|$. As shown in (2), the α_{ij} lead directly to the time evolving state vector $|\psi(t_i)\rangle_R$. One uses the information from the state vector at each time step to infer information about H. Together (4) and (5) give a total of N_R complex (or $2N_R$ real) constraints on H. Since a general $N_R \times N_R$ Hamiltonian has N_H^2 real degrees of freedom, the α_{ij}'s do not contain enough information to define a full Hamiltonian at each time step. After all, the α_{ij}'s only tell us about the evolution of a single state, whereas the Hamiltonian contains full information about the evolution of all possible states.

The fact that the Hamiltonian is highly underdetermined by a single time step can be exploited to add the condition that the Hamiltonian is time independent without loss of generality. As long as one is looking at no more than $N_H/2$ time steps, (4) and (5) provide no more than N_H^2 real constraints which can be used to build at least one time independent Hamiltonian that describes the full time evolution. And to the extent that the α_{ij} are randomly generated, the Hamiltonians produced from the α_{ij}'s should be randomly distributed as well. In fact, it seems reasonable to expect that the central limit theorem will give the distribution of Hamiltonians (generated by effectively inverting (4) and (5)) an enhanced degree of Gaussianity over whatever distribution generated the α_{ij}'s.

For all this to work out, we need to constrain the number of time steps N_t according to

$$N_t < N_H/2. \tag{6}$$

We can estimate N_t as the age of the Universe divided by the minimum time resolution δt. Using arguments from Sect. 4, $\delta t \equiv 1/\Delta E$ and the maximum value of ΔE ($= 10^{11}$ GeV) gives

$$N_t \approx \frac{\Delta E}{H_0} = \frac{10^{11}\text{GeV}}{10^{-42}\text{GeV}} = 10^{53}. \tag{7}$$

By comparison, requiring a good match of the density of states to a field theory leads to (11) giving

$$N_H \geq \frac{B}{a}\frac{E_M}{E_0}\left[1-\left(\frac{E_0 - E_S}{E_M}\right)^{\beta}\right]^{-\gamma}\exp\left[b\left(\frac{E_0}{\Delta k}\right)^{\alpha}\right] \tag{8}$$

The quantity $E_0/\Delta k$ in the exponent is the ratio of the energy of the Universe to the field theory k-space cutoff. Even choosing values from Sect. 4 which minimize the bound on N_H gives exponentially large values for the *exponent* in (8) and gives lower bounds on N_H which easily satisfy (6) and validate the assumption of a time independent Hamiltonian.[1]

4 Field Theory and the Wigner Semicircle

4.1 Our Basic Approach

The clock ambiguity implies that any random split of superspace into clock and rest subsystems should lead to a realization of "physical laws". However, one expects that a random split would result in laws described by a random Hamiltonian. In [4] we discussed possible ways forward under those conditions. One thing we did was pose the question in the converse form to test this hypothesis. Namely, we evaluated the extent to which the known physical laws match to those derived from a random Hamiltonian evolution. In particular, we compared the spectrum of a free field theory, representing (approximately) the known physics, to the eigenvalue spectrum of random Hamiltonians.

Following [4], we do not undertake the project of specifically constructing field operators etc. in terms of the eigenstates of the Hamiltonian. This project is likely to be challenging, and is also likely to further involve a statistical analysis of different

[1] This argument appears to be very robust. For example, refining the time resolution to $\delta t = 1/M_P$ does not change the result at all.

physical realizations consistent with the same eigenvalue spectrum and initial state $|\psi(t_1)\rangle_R$. We feel that our analysis at the level of the eigenvalue spectrum represents a first check of the viability of our line of reasoning, and we save the important question of defining field operators etc. for future stages of this work.[2]

The distribution of eigenvalues for a random Hamiltonian, represented as an $N_H \times N_H$ Hermitian matrix, follows under quite general assumptions [8] the Wigner semicircle rule in the large N_H limit. Take, for example, the distribution of eigenvalues of a large Hermitian matrix with elements drawn from a Gaussian distribution depicted in Fig. 2.

On the other hand, the density of states for a free field theory grows, at large energies, like an exponential of a power of the energy. On the face of it, these two forms for dN/dE are dramatically different. In order to press forward with the comparison we introduced a general parametrization for the random Hamiltonian and field theory spectral densities respectively:

$$\frac{dN_R}{dE} = \begin{cases} a\frac{N_H}{E_M}\left(1-\left(\frac{E-E_S}{E_M}\right)^{\beta}\right)^{\gamma} & |E-E_S| < E_M, \\ 0 & \text{otherwise}, \end{cases} \tag{9}$$

$$\frac{dN_F}{dE} = \frac{B}{E}\exp\left\{b\left(\frac{E}{\Delta k}\right)^{\alpha}\right\}, \tag{10}$$

where E_M and E_S represent the maximum eigenvalue of the random Hamiltonian and an offset energy between the two descriptions, Δk ($\equiv 2\pi/L$) is the resolution in k-space set by putting the field theory in a box of size L and B, b, α and γ are dimensionless parameters.

Expanding both (9) and (10) in a Taylor series around a given central energy $E_0 = \rho R_H^3 = 10^{80}\text{GeV}$, corresponding to the current energy of the Universe, and trying to equate the results at each order in $(E - E_0)$ we find the level of agreement between the two descriptions.

Equating the zeroth order sets the size of the space of the random Hamiltonian to be exponentially large:

$$N_H = \frac{B}{a}\frac{E_M}{E_0}\left[1-\left(\frac{E_0-E_S}{E_M}\right)^{\beta}\right]^{-\gamma}\exp\left[b\left(\frac{E_0}{\Delta k}\right)^{\alpha}\right]. \tag{11}$$

Strictly speaking, this expression only gives a lower bound on N_H, since we only really know upper bounds on Δk.

Equating the first order (as well assuming equality at zeroth order) sets the offset energy E_S in terms of the energy of the Universe E_0 by the following implicit expression:

[2]Lee Smolin has drawn our attention to work by Bennett et al. [7] which may offer a framework where specific symmetries and representations for elementary particles could be predicted in a scheme such as ours.

$$-\beta\gamma\frac{E_0}{E_0-E_S}\frac{\left(\frac{E_0-E_S}{E_M}\right)^\beta}{1-\left(\frac{E_0-E_S}{E_M}\right)^\beta}=\alpha b\left(\frac{E_0}{\Delta k}\right)^\alpha. \tag{12}$$

Assuming equality at zeroth and first order, the relative difference at the second order is fixed and given by

$$\Delta_2\equiv\frac{\Delta\frac{dN}{dE}}{\frac{dN}{dE}|_{E_0}}\approx\frac{\alpha^2b^2}{\gamma}\left(\frac{E_0}{\Delta k}\right)^{2\alpha}\frac{(E-E_0)^2}{E_0^2}. \tag{13}$$

Table 1 shows the value of Δ_2 for different values of the exponent α in (10), the field theory k-space lattice spacing Δk and the range of validity of the field theoretical description

$$\Delta E = E - E_0 \tag{14}$$

which can be thought of in terms of a minimum timescale on which field theory is valid, given by $\delta t \sim 1/\Delta E$. The idea is to check if the disagreement between the density of states of a random Hamiltonian and a free field theory at second order, Δ_2 can be "sufficiently small" for realistic parameters. We find that the parameter most critical to this analysis is α, and we discuss its value in the next section.

4.2 The Value of α and the Dimensions of Space

The results for the density of states of a field theory in $1+1$ dimensions for bosons and fermions can be derived from different instances of the Cardy formula for conformal field theories in 2d [9]. This formula relates the entropy of the field theory to its energy E and central charge c in the following way

$$S=\log N(E)=\frac{1}{2\pi}\sqrt{\frac{c}{6}\left(E-\frac{c}{24}\right)}, \tag{15}$$

and implies (10) with exponent $\alpha = 1/2$. The asymptotic density of states can also be found for a conformal field theory in higher number of dimensions [10] and grows as $e^{E_E^{(d-1)/d}}$ where E_E is the extensive energy. However, if the Casimir energy E_C is taken into account the total energy $E = E_E + E_C$ is sub-extensive and the dependence of the entropy on energy changes. Verlinde [11], based on holographic arguments, proposed that the Cardy formula is satisfied also in the case of higher dimensional field theories.

Taking the extensive energy expression for a field theory in $3+1$ dimensions would fix the constant $\alpha = 3/4$ in our parametrization of the density of states (10). A first assessment of Table 1 indicates that the agreement between the field theory and random Hamiltonian would be poor (with $\alpha = 3/4$, $\Delta_2 \gg 1$ for all entries). An alternative interpretation might be to note that the transition from $\alpha = 1/2$ to

Table 1 Value of Δ_2 for different choices of α, Δk and ΔE. As in [4], values for ΔE are set by accelerator (10^3GeV) or cosmic ray (10^{11}GeV) bounds. Values for Δk are set by the photon mass bound (10^{-25}GeV) or the size of the Universe (10^{-42}GeV)

α	Δk (GeV)	ΔE (GeV)	Δ_2
1/2	10^{-25}	10^3	$10^{-24.5}$
1/2	10^{-25}	10^{11}	$10^{-16.5}$
1/2	10^{-42}	10^3	10^{-16}
1/2	10^{-42}	10^{11}	10^{-8}
3/4	10^{-25}	10^3	$10^{1.8}$
3/4	10^{-25}	10^{11}	$10^{9.8}$
3/4	10^{-42}	10^3	$10^{14.5}$
3/4	10^{-42}	10^{11}	$10^{22.5}$
1	10^{-25}	10^3	10^{28}
1	10^{-25}	10^{11}	10^{36}
1	10^{-42}	10^3	10^{45}
1	10^{-42}	10^{11}	10^{53}

$\alpha = 3/4$ in our table occurs roughly right at the point where Δ_2 shifts from small to large values. Given that all our estimates are very rough at this stage, there may be a hint here of a way in which our methods could *predict* the three dimensions of space, as the maximum value consistent with a random Hamiltonian.

On the other hand, if we assume Verlinde is correct and use the universal Cardy formula, that implies $\alpha = 1/2$ for any d. Then the difference Δ_2 is negligible and random Hamiltonians give a density of states that appears strongly consistent with the field theoretical one, at the expense of any apparent preference for the value of d.

4.3 *Wigner's Tail*

It may appear disturbing that we are attempting to match expressions (9) and (10), the latter having positive second derivative everywhere while the former in the case of the Wigner semicircle is negative definite; the case depicted in Fig. 1. As discussed above, it may simply be the case that this discrepancy is negligible, and is not a problem.

One might also wonder if this may change if the perfectly Gaussian probability distribution is altered, for example, if the width of the distribution of eigenvalues is different in different energy ranges.[3] To be concrete, one may consider the distribution containing a small cubic piece. In such a case (studied in [12]) the exponent γ in the density of states may be changed from $1/2$ (Wigner semicircle) to $3/2$ which has regions of positive second derivative near the tails of the distribution as depicted in Fig. 2. This possibility is included in our parametrization given in (9).

[3] We thank Jaume Garriga for suggesting this direction of investigation.

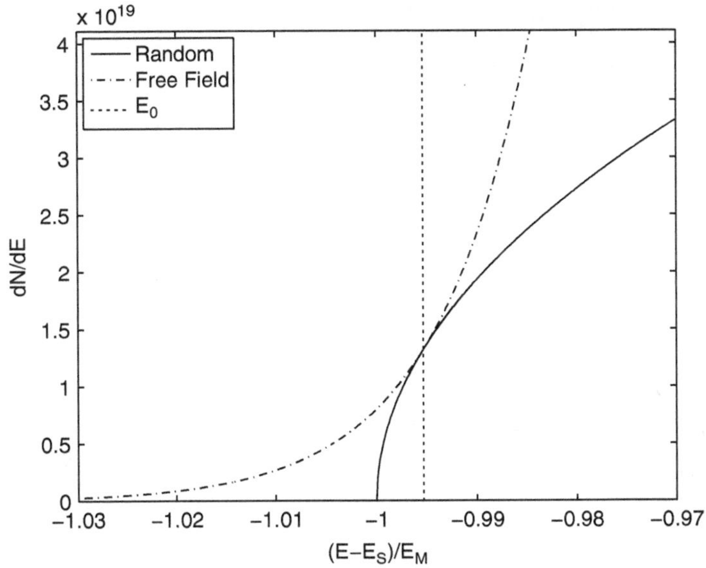

Fig. 1 A plot of the density of eigenvalues for a random Hamiltonian using (9) and a field theory using (10) matching the zeroth and first order terms in a Taylor expansion around E_0 (the *vertical line*).

The corresponding improvement in matching can be inferred from (13); an increase in γ leads to a smaller relative difference Δ_2.

Let us point out, as a curiosity, that a distribution highly distorted from Gaussianity might lead to a perfect matching with the field theory distribution. In fact, letting γ grow makes the generalized random density of states (9) approach an exponential of the form of the field theory one (10). In order to see this we may take $E_M \gg E_0 - E_S$ in (12) to find

$$-\gamma\left(\frac{E_0 - E_S}{E_M}\right)^{\beta} \approx \left[\frac{\alpha}{\beta}\left(1 - \frac{E_S}{E_0}\right)\right] b\left(\frac{E_0}{\Delta k}\right)^{\alpha}, \tag{16}$$

and choose parameters such that the coefficient in brackets is approximately one. Therefore, we have that the random density of states has the form

$$\frac{dN_R}{dE} = a\frac{N}{E_M}\left(1 + \frac{x}{\gamma}\right)^{\gamma}, \tag{17}$$

where, $x \equiv -\gamma(E - E_S/E_M)^{\beta} \approx b(E/\Delta k)^{\alpha}$ for $E \approx E_0$, that in the limit of large γ reproduces the exponential behavior of the field theory density of states. However, we don't think that such a distortion of the distribution could be the outcome of a truly statistical averaging procedure. Furthermore, it seems contradictory to the spirit of this work to seek out an exotic distribution. That would appear to undermine the hope that our methods could one day offer some real predictive power.

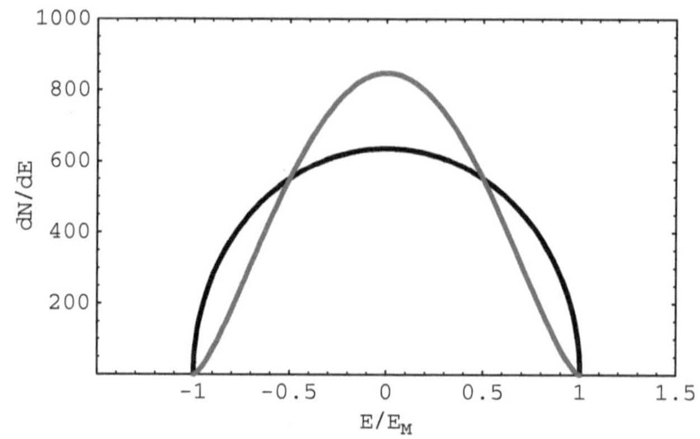

Fig. 2 A plot of the density of eigenvalues for a random Hamiltonian ($E_M = 1$, $N_H = 1000$) in the cases of: a Gaussian distribution (*black*) giving rise to the Wigner Semicircle and Gaussian plus a cubic "interaction" term (*gray*) with concave tails.

5 Including Gravity

In this and previous work we have not discussed gravity at length. In [4] we suggested that gravity could naturally emerge when a more general metric is allowed when interpreting a random Hamiltonian as a local field theory (vs. the Minkowski metric implicit in the discussion in Sect. 4). In such a picture we do not expect a full consistent theory describing arbitrary spacetimes to emerge. It would be enough to get a theory of spacetime that would be consistent for the actual state of the Universe and similar states. It is not even clear, for example, that the full number of states associated with black hole entropy would need to be part of the spectrum in such a picture, since the microscopic properties of black holes do not really appear to be part of our physical world . It seems reasonable to proceed carefully with this issue, and avoid jumping to conclusions about gravity in this picture until some of these ideas have been worked out more systematically.[4]

In the next section we will try a different approach. Specifically, we will relate the curvature of the Wigner semicircle to the specific heat of the Universe. In estimating the specific heat we use standard notions of the heat capacity of gravitating systems, and thereby implicitly introduce gravity into our analysis. We do this with the caveat that this approach may take us even further out on a limb than the other (admittedly speculative) ideas discussed elsewhere in this paper. Interestingly, the analysis in the next section yields intriguing results even when the more exotic forms of gravitational entropy (black hole and De Sitter entropy) are ignored. Thus the

[4]We find it intriguing that this picture bears some resemblance to approaches that explicitly reject a full "third quantized" superspace formalism, such as that discussed in [13].

analyisis of Sect. 6 seems to apply even in the context of the more conservative ideas about gravity reviewed in this section.

6 Heat Capacity and N'''

Here we return to curvature of the dN/dE vs E curve, i.e., the third derivative of $N(E)$, and estimate its value from a thermodynamic perspective. We will use the fact that the heat capacity (or its intensive counterpart, the specific heat) is a thermodynamic quantity related to N'''. As discussed in the previous section, we will incorporate gravity by considering thermodynamic quantities defined for gravitating systems such as black holes.

Our starting point is the standard canonical ensemble expression for the entropy of a system with energy in a range ΔE around a central energy E_0:

$$S = \log\left(\frac{dN(E_0)}{dE}\Delta E\right) \tag{18}$$

This leads to

$$\frac{1}{T} \equiv \frac{dS}{dE} = \frac{d(\log(\frac{dN}{dE}\Delta E))}{dE} = \frac{N''}{N'}, \tag{19}$$

and using $C \equiv dE/dT$

$$\frac{1}{C} = \frac{d}{dE}\left(\frac{N'}{N''}\right) = 1 - \frac{N'N'''}{N''^2}. \tag{20}$$

When discussing these thermodynamic quantities one must generally be careful to state what is being varied and what is being held fixed when differentiating. We will return to that question a bit later in this section.

Plugging the generalized Wigner form (9) for the density of states into (20) gives

$$\frac{1}{C} = 1 - \frac{(\gamma - 1)Q - (\beta - 1)}{2\beta^2\gamma Q}. \tag{21}$$

Here $1/Q = ((E_0 - E_S)/E_M)^\beta - 1$ is an exponentially small quantity if the order by order matching described in Sect. 4.1 is performed. Thus, to an excellent approximation we have

$$C = \left(1 + \frac{1-\gamma}{2\beta^2\gamma}\right)^{-1}. \tag{22}$$

Taking parameters around the Wigner case ($\gamma = 1/2$ and $\beta = 2$) gives $C = 9/8$.

Originally, the motivation for this line of investigation was the following: The second derivative of the density of states of the Wigner semicircle has the opposite sign to that of the field theory density of states (as can be seen by inspecting Fig. 1). The heat capacity is related to the second derivative of the density of states, and is negative for strongly gravitating systems. Strongly gravitating systems dominate the entropy of the Universe, so perhaps the negative specific heat of strongly gravitating systems in the Universe allows one to more fully reconcile the density of states of real matter with the Wigner semicircle at second order. This idea is not realized however, because the the second derivative of the density of states is not related to the specific heat in a sufficiently simple way. For the cases we consider, the second derivative of the density of states remains positive, even when the specific heat is negative. Forced to abandon this simple idea, we none the less move forward with the comparison with thermodynamic quantities which still turns out to give interesting results.

Due to the additivity of the entropy, it will be convenient to work with the derivatives of entropy with respect to E

$$d^n S/dE^n. \tag{23}$$

These quantities can be constructed by differentiating $S(E)$ directly, or they can be constructed from other thermodynamic quantities. For example, the $n = 2$ case can be related to the heat capacity using

$$\frac{1}{T^2 C} = -\frac{d^2 S}{dE^2} \tag{24}$$

(which can be derived from Eqns. 19 and 20).

If we write the entropy of the Universe as a sum over different components (such as radiation and black holes) labeled by i one has

$$\frac{d^2 S_{tot}}{dE^2} = \frac{d^2}{dE^2}\sum_i S_i = \sum_i S_i'' = -\sum_i \frac{1}{T_i^2 C_i}\,. \tag{25}$$

The Wigner density of states (Eqn. 9) gives

$$-\frac{d^2 S}{dE^2} = \frac{N''^{\,2} - N' N'''}{N'^{\,2}} = \frac{1+\beta Q}{E - E_S}\frac{dS}{dE} = \gamma\beta\frac{(1+Q)(1+\beta Q)}{(E-E_S)^2}\,. \tag{26}$$

We wish to compare Eqn. 26 with Eqn. 25. To do so we will either estimate T_i and C_i or S_i'' directly for the various components of the Universe. We consider four main contributions coming from radiation (R), black holes (BH), dark matter (DM) and dark energy (DE).

Radiation: To compute the radiation component we take a gas of photons with energy $E_R = \rho_R H^{-3} = T_R^4 H^{-3}$ and temperature $T_R = 10^{-13} GeV$. Keeping the volume $H^{-3} = (10^{-42} GeV)^{-3}$ fixed we obtain

$$C_R = 2 \times 10^{88} , \qquad \frac{1}{T_R^2 C_R} = 10^{-62} GeV^{-2} , \tag{27}$$

and entropy $S_R = 4C_R/3 \sim 10^{88}$.

Black Holes: We use the total black hole entropy estimate of [14]:

$$S_{BH}^{tot} = \sum_{N_{gal}} 4\pi \frac{M_{BH}^2(gal)}{m_{pl}^2} \sim 3.2 \times 10^{101} \frac{E_{BH}}{10^{75} GeV} \left(\frac{M}{10^7 M_{\odot}} \right) , \tag{28}$$

where the sum is over galaxies ($N_{gal} \sim 10^{11}$) within the volume H^{-3} and $M_{BH}(gal)$ is the distribution of masses of supermassive black holes at the galactic cores, which we approximate here as being peaked at $M = 10^7 M_{\odot}$. Using $E_{BH} = N_{gal} M = 10^{11} 10^7 M_{\odot} = 10^{-5} E_0$ and $M_{\odot} = 10^{57} GeV$ (*i.e.*, $T_{BH} \sim 10^{64} GeV$) we obtain

$$C_{BH} = -2.1 \times 10^{91} \left(\frac{M}{10^7 M_{\odot}} \right)^2 , \qquad \frac{1}{T_{BH}^2 C_{BH}} \sim -10^{-38} GeV^{-2} . \tag{29}$$

Dark Matter: We infer the dark matter temperature by equating the dark matter kinetic energy with thermal energy:

$$T_{DM} \sim \left(\frac{v}{100 km/s} \right)^2 \frac{m_{DM}}{100 GeV} 10^{-4} GeV \sim 10^{-4} GeV , \tag{30}$$

with m_{DM} being the mass of the dark matter particle. We consider that only a fraction of the energy differential dE, of order $v^2/c^2 \sim 10^{-3}$, goes into thermal energy. These leads to a dark matter heat capacity of order

$$C_{DM} \sim \pm 10^{-6} , \qquad \frac{1}{T_{DM}^2 C_{DM}} \sim \pm 10^{-2} GeV^{-2} . \tag{31}$$

In virialized bound systems there would be a negative contribution coming from the gravitational energy twice as large as the kinetic component leading to a negative heat capacity (and the negative sign in Eqn. 31). Non-bound dark matter would contribute with a positive sign. We allow both signs in Eqn. 31 because our analysis is not detailed enough to consider which effect dominates.

Dark Energy: We use the de Sitter entropy $S_{DE} = E^2/m_{Pl}^2 \sim 10^{120}$ (with $E \equiv \rho_{DE} H^{-3}$) giving

$$d^2 S_{DE}/dE^2 \sim 2/m_{Pl}^2 \sim 10^{-40} GeV^{-2} \tag{32}$$

with a temperature of order $T_{DE} \sim H_0 \sim 10^{-42} GeV$.

Total for the Universe: Because the Universe is comprised of different components which are not in equilibrium, we work with Eqn. 24 which is easy to treat as a sum of independent components. Plugging all four components into Eqn. 24 (with $i = \{R, BH, DM, DE\}$) leads to an expression of the form

$$\frac{1}{T^2 C} = -\sum_i \frac{d^2 S_i}{dE^2}, \tag{33}$$

to be compared with $(1 + \beta Q) dS/(E_0 - E_s) dE$ (from Eqn. 26) for the random Hamiltonian.

We notice that the ratios of S_i to $dS_i/dE = T_i^{-1}$ and of dS_i/dE to $d^2 S_i/dE^2 = T_i^{-2} C_i^{-1}$ for each component in the above estimates are all of order E_0. The regularity of these ratios makes it possible to reconcile the two descriptions if the following relation holds:

$$\frac{1 + \beta Q}{E_S - E_0} \sim \frac{1}{E_0}, \tag{34}$$

which at this point of our analysis does not lead to any inconsistency with our previous results since the parameter E_S was still unconstrained.

Indeed, the generalized distributions we proposed, Eqns. (9) and (10), have more free parameters than constraining equations, Eqns.(11)-(13), even after setting $\alpha = 1/2$. Therefore, it appears that demanding consistency as we have done above does not produce onerous constrains on the system. A caveat to this conclusion could come from any insights that suggest that the properties of ratios of derivatives scaling as E_0^{-1} is non-trivial for the actual Universe, but on the face of it this seems to be a straightforward result that obtains for a great variety of functional forms for $S(E)$.

An interesting feature of the above discussion is that it applies to a variety of different cases: The entropy and its various derivatives calculated above are clearly dominated by the contributions from S_Λ. But one could "conservatively" argue that S_Λ is quite abstractly defined, and should not be allowed to contribute to comparisons with the Wigner density of states. Perhaps the Wigner density of states should only be equated with degrees of freedom that are more physically observable. Removing S_Λ from the computation would allow S_{BH} to dominate. Since ratios of derivatives of S_{BH} have the same properties, the comparison with Wigner goes through unchanged. Similar arguments might cause one to leave out S_{BH} as well. Then S_{DM} dominates and again the analysis goes through.

Interestingly, if one considers the dark matter to be dominant, one can consider integrating the discussion here with the comparison of Wigner with field theory in Minkowski space in Sections 4.1 and 4.2. The possibility that most of the dark matter entropy is in states that are only linearly perturbed gravitationally is consistent with current observations, and under those conditions it may be reasonable to *combine* the constraints presented here with those from Section 4. The value of E_S needed to satisfy Eqn. (34) together with the field theory requirements is exponentially close to $-E_M$, half the width of the Wigner distribution, with $E_0 \ll E_M$.

What are we to make of this comparison? We are trying to learn if the Wigner semicircle gives a sufficiently good approximation to the density of states of the Universe. Our current analysis assumes that it is possible to take the Wigner semicircle density of states in the vicinity of some energy E_0 and set up a correspondence with eigenstates of a Hamiltonian that describes the Universe more or less as we know it. In this section we assume this correspondence allows us to use the thermodynamic quantities as estimated above. Specifically, the differentiation with respect to E should reflect the differences between the thermodynamic quantities calculated at E_0, and for a similar cosmological interpretation of the Wigner density of states an energy dE away. A careful understanding of how the black holes, radiation, etc. change as one shifts by dE and reinterprets the density of states cosmologically would be required to give our calculations more rigor (of the sort commonly expressed, for example, by holding specified properties fixed when differentiating thermodynamic quantities). In the absence of such rigor, we hope that the simple differentiations performed in this section give a reasonable approximation to the desired result.

The crudeness of our methods warrant a great deal of caution, but we still find it a curiosity, perhaps even an encouraging curiosity that our comparison yields results that are comparable within an order of magnitude, and possibly even with the right sign.

7 Summary and Conclusions

The clock ambiguity suggests that we must view physical laws as emergent from a random ensemble of all possible laws. We started this article with a review of our earlier work showing the origin of the clock ambiguity. We then outlined and expanded upon our earlier ideas about the central role of quasi-separability in such a statistical analysis, and discussed how this could lead to a prediction that local field theory should provide the basic form for physical laws. We have shown that (contrary to our earlier assumptions) one can quite generally assume physical laws that emerge in this picture will have a time independent Hamiltonian. We reviewed our earlier work that shows how the density of states of a free field theory can be well approximated by a random Hamiltonian, and extended this work to include a possible predictive link to the number of dimensions of space. We also explored a higher order analysis that (favorably) compares the curvature of the density of states of a random Hamiltonian with that of the observed Universe using estimates of the specific heat of the various components of the Universe.

While most of our discussion here is rather heuristic, our new results all add to the case that a statistical approach to physical laws may indeed be viable. In the case of the time independence of the Hamiltonian, we feel we have presented a very solid result which gives a significant improvement over our earlier discussions. All in all, while many open questions remain that could ultimately undermine our approach, we feel that a statistical approach to the emergence of physical laws remains an interesting possibility which has accumulated additional support from the work presented here.

Acknowledgements We would like to thank T. Banks, S. Deser, R. Emparan, B. Fiol, J. Garriga, L. Knox, H. Nielsen, L. Smolin, G. Starkman, A. Tyson and S. White for valuable discussions. This work was supported in part by DOE Grant DE-FG03-91ER40674.

References

1. R. Arnowitt, S. Deser, C.W. Misner, Phys. Rev. **117**, 1595–1602 (1960)
2. C.J. Isham (1992) [gr-qc/9210011]
3. A. Albrecht, Lect. Notes Phys. **455**, 323–332 (1995) [gr-qc/9408023]
4. A. Albrecht, A. Iglesias, Phys. Rev. **D77**, 063506 (2008) [arXiv:0708.2743[hep-th]]
5. C. Wetterich, Nucl. Phys. **B314**, 40 (1989)
6. C. Wetterich, Nucl. Phys. **B397**, 299-338 (1993)
7. D. Bennett, N. Brene, H. Nielsen, Phys. Scr. **T15**, 158 (1987)
8. M.L. Mehta, *Random Matrices* (Academic Press, Boston, 1991)
9. J.L. Cardy, Nucl. Phys. **B270**, 186–204 (1986)
10. T. Banks (1999) [hep-th/9911068]
11. E.P. Verlinde (2000) [hep-th/0008140]
12. E. Brezin, C. Itzykson, G. Parisi, J.B. Zuber, Commun. Math. Phys. **59**, 35 (1978)
13. T. Banks, W. Fischler, S. Paban, J. High Energ. Phys. **12**, 062 (2002) [hep-th/0210160]
14. T.W. Kephart, Y.J. Ng, J. Cosmology Astropart. Phys. **0311**, 011 (2003) [gr-qc/0204081]

Holographic Cosmology and the Arrow of Time

Tom Banks

Abstract I review the holographic theory of quantum space-time and cosmology, and argue that it may yield insight into the Boltzmann-Penrose question of why the universe began with low entropy. In this model, the observed low entropy initial conditions, *may* be the most general initial conditions for cosmology, which avoid collapse into a dense black hole fluid phase.

1 General Relativity Is Not a Quantum Field Theory!

The basic paradigm of quantum field theory in a fixed space-time is based on the Cauchy–Kowalevskaya theorem for second order hyperbolic partial differential equations. This theorem states that, given smooth values of the fields and their first time derivatives on a complete space-like slice (a Cauchy surface), there exists a solution of the equations in a finite region of the space-time, foliated by Cauchy surfaces to the future of the initial value surface. Thus, the phase space, or space of solutions of the equations, is parametrized by the values of canonical pairs of variables on the initial value surface. One quantizes by specifying the operator algebra of the canonical variables on the initial value surface. For a smooth fixed background metric, and a renormalizable Lagrangian, the quantum theory can be constructed in a manner covariant under infinitesimal coordinate transformations.

An analogous formulation of classical GR was given by Arnowitt Deser and Misner (ADM). Formal quantization of it leads to the Wheeler–DeWitt equation. It is the basis of the loop quantum gravity program. There are numerous problems with

T. Banks (✉)
Department of Physics and SCIPP, University of California, Santa Cruz, CA 95064, USA

Department of Physics and NHETC, Rutgers University, Piscataway, NJ 08540, USA
e-mail: banks@scipp.ucsc.edu

L. Mersini-Houghton and R. Vaas (eds.), *The Arrows of Time*, Fundamental Theories of Physics 172, DOI 10.1007/978-3-642-23259-6_5,

this approach,[1] but I would like to emphasize one in particular. Above three space-time dimensions (and in three spacetime dimensions with negative cosmological constant (c.c.)) generic initial data lead to singular solutions.

The field theory paradigm leads to the slogan "one degree of freedom per space point" on the initial value surface. This idea is problematic from the start when the formalism tells us that the volume of space associated with a given coordinate region can change with time. However, when entire regions of space can be swallowed up inside singularities, we should clearly be thinking more carefully about precisely what the actual phase space is that we are trying to quantize. There are precious few global existence theorems in GR, and those that exist [1] obviously put stronger restrictions on initial data than one would want. The point is that we believe that some of the singular solutions, namely black holes, actually represent physical processes which *can* occur in the real world, and should be included in the theory.

The correct conjecture to make is called Cosmic Censorship: with some restrictions on initial data, the only types of singularities that occur are black holes.[2] In formal GR jargon: singularities should always be in the complement of the causal past of future infinity. Of course, this may not include all possible space-times that we would like to understand: it legislates away both Big Bang and Big Crunch singularities. However, Cosmic Censorship gives us a handle on a class of space-times which we can study reliably, and therefore gain insight into the nature of the degrees of freedom of quantum gravity.

The Cosmic Censorship conjecture has never been proved. Part of the reason for this is that one is not sure exactly how to formulate the restriction on initial conditions, which would guarantee it to be true. I'd like to make a suggestion in this connection. String theory has shown us that (in asymptotically flat space-time of dimension higher than four), the gauge invariant observables are all encoded in the S-matrix for finite numbers of incoming and outgoing stable particles. It would seem then that the correct initial conditions for the Cosmic Censorship conjecture in

[1]In two dimensions, where the formalism actually leads to a well defined mathematical construction, one is forced to include processes in which the universe splits into disconnected pieces, in order to describe a unitary quantum theory. One obtains the topological expansion of string theory, which only makes sense for a very restricted class of generally covariant second Lagrangians. The observables of these well defined models have no obvious connection to local measurements on a two dimensional world sheet. Similar remarks are valid in one space-time dimension, though the freedom to choose the world-line Lagrangian is not compromised in this case.

[2]There is a possible modification of this suggested by string theory. We have evidence in string theory for the existence of naked singularities like orbifolds, orientifolds, conifolds, which are resolved by the quantum theory. For example, singular limits of $K3$ manifolds are perfectly regular string compactifications. One can certainly imagine that one can start with a classically regular string compactification on $K3$ and choose initial conditions for the moduli, which send them through a singular point in moduli space, in some localized region in the non-compact dimensions. It is possible, that this does not give rise to a black hole, but simply that quantum effects become important, as new light states have to be included in the effective field theory description of physics near the singularity.

asymptotically flat space-time should be incoming scattering data: a finite number of incoming localized wave packets of arbitrarily small amplitude, but arbitrarily high energy. In four dimensions, one would have to understand the Fadeev–Kulish [2] generalization of the S-matrix to include coherent bremstrahlung radiation of massless particles. One should also be aware of possible exceptions to the theorem involving resolvable naked singularities, as noted in the previous footnote.

Similarly, Cosmic Censorship in Asymptotically AdS space-time should assume initial conditions which are sums of infinitesimal waves, satisfying the usual normalizable boundary conditions [3]. We have plenty of evidence [4] that boundary conditions only slightly different than these, can cause singularities, and that they do not correspond to states in the same quantum theory.

This analysis of the phase space of classical GR suggests that the key to understanding the local nature of the degrees of freedom of the quantum theory, is to understand black holes. There are other lines of argument, which lead to the same conclusion. Since Rutherford, we have become used to searching for the "ultimate building blocks" by performing scattering experiments in a regime where all kinematic invariants are large. The work of [5] is evidence that, once the impact parameter becomes smaller than the Schwarzschild radius corresponding to the center of mass energy, such scattering processes produce singularities, which are black holes if Cosmic Censorship is correct.

The Bekenstein–Hawking formula for black hole entropy, also suggests that the ultimate high energy behavior of theories of quantum gravity is dominated by black holes. The validity of the AdS/CFT description of asymptotically AdS quantum gravity depends on the fact that this formula matches that of the boundary CFT. This can be viewed as rigorous evidence for black hole dominance of high energies, if we take the CFT as the definition of quantum gravity in asymptotically AdS space-time. In asymptotically flat space, the black hole entropy formula grows much more rapidly than any quantum field theory or perturbative string theory. However, as we will see below, for certain dimensions it is consistent with a field theory description of the light-cone Hamiltonian.[3]

I would also remind the reader of the lesson taught us by Feynman and Wilson: the degrees of freedom we use to describe all known quantum systems, are those which dominate the high energy behavior. Indeed, the Wilsonian definition of a quantum field theory (which includes all standard quantum systems with infinite dimensional Hilbert spaces as special cases) is that it is a relevant deformation of the conformal field theory which dominates its high energy behavior.

These observations originally led me to formulate the idea of *asymptotic darkness*[6]. Originally, I had imagined that one could access a theory of quantum gravity by starting from black hole states, treated as stable and exactly degenerate, and use what we know about semi-classical black hole physics to figure out a consistent theory. The apparent problem is that black holes are thermal systems,

[3]That is, the light cone Hamiltonian *may* be identified with a time-like generator in an ordinary Lorentz invariant field theory.

and semi-classical physics tells us only about their thermodynamic properties. These depend only on rather generic features of the Hamiltonian. The idea of asymptotic darkness *is* of some utility, as demonstrated by the following examples:

- In d dimensional Anti-de Sitter space (AdS_d), with radius R, the asymptotic Bekenstein–Hawking formula for black hole entropy reads

$$S(E) = c_d(R^2 E M_P)^{\frac{d-2}{d-1}}.$$

  c_d is a numerical constant. This is identical to the entropy formula of a $d-1$ dimensional conformal field theory, compactified on a spatial sphere of radius R The "number of degrees of freedom" of the CFT is related to the AdS radius in Planck units. This observation could have been the genesis of the AdS/CFT correspondence.
- In asymptotically flat space, the BH formula grows more rapidly that E, and cannot be matched by any quantum field theory. However, the entropy as a function of light-front energy for $d > 4$ grows less rapidly than E[7]. For $d > 6$, Matrix Theory gives us a model of the light front Hamiltonian of theories with ≥ 16 supercharges as a limit of quantum field theories with $N \times N$ matrix variables. The finite matrix field theories are a Discrete Light Cone quantization of Super-Poincaré invariant quantum gravity, and have a density of states which grows too rapidly at infinite energy, to match to the BH formula. Indeed, the quantum gravity states are part of the *low* ($o(1/N)$) energy spectrum of the field theory. The DLCQ theory is not a field theory for $d = 5, 6$, but the BH entropy formula suggests that the $N \to \infty$ limit might be a $9-d$ dimensional field theory.
- The BH entropy formula suggests that theories with linear dilaton asymptotics have a Hagedorn spectrum. This is consistent with the idea that these are Little String Theories [8], in which the bulk gravitational constant is dialed to zero, so that perturbative estimates of the high energy density of states are valid. It also suggests a formulation of the light cone Hamiltonian of these systems in terms of $1+1$ dimensional field theory. This is indeed correct for the DLCQ of Little String Theory [9].

These examples also demonstrate a fundamental property of quantum gravity, which distinguishes it sharply from quantum field theory: the high energy behavior of the Hamiltonian, and its very definition, depend on the *infrared* space-time asymptotics. Indeed, for general space-time asymptotics we should not expect a time-independent Hamiltonian at all. How then are we to generalize these insights? I believe that the fundamental clue lies in the covariant entropy bound [10]. This relates entropy not to a specialized concept like conserved energy, but to the general, and local, concept of *the area of a holographic screen.*

A causal diamond in a Lorentzian space-time is the intersection of the interior of the backward light cone of a point Q and that of the forward light cone of a point P in the causal past of Q. For sufficiently small time-like separation between

P and Q, the area of any $d-2$ surface on the null-boundary of the diamond is finite, and there is a unique maximal area $d-2$ surface on the boundary, called the holographic screen of the diamond.[4] The covariant entropy bound asserts that the entropy which can be accessed by observers in the interior of the diamond is bounded by one quarter of this maximal area, in Planck units.

In quantum field theory, one associates an infinite dimensional algebra of observables[5] with each causal diamond. The covariant entropy bound thus implies, as we have remarked above, that quantum gravity is not a quantum field theory. Strictly speaking, this follows from a slightly stronger statement, formulated by Fischler and myself [11], to the effect that the entropy referred to in the bound is the entropy of the maximally uncertain density matrix associated with measurements made in the causal diamond. In general, as we have noted, gravity does not allow us to define special operators like the energy for a finite causal diamond, which could give sense to the concept of canonical, or micro-canonical density matrices. If the covariant entropy bound is indeed a general feature of quantum gravity, then it seems reasonable that the covariant entropy bound refers to the maximally uncertain density matrix on the Hilbert space. In particular, the finite entropy of a causal diamond, implied by the bound, is the statement that the algebra of observables associated with that diamond in a quantum theory of gravity is finite dimensional.[6]

This remark is exciting for several reasons:

- It gives a purely quantum mechanical definition of a quantity which is a geometrical area, in the limit that that quantity is large.
- Similarly, we can give a purely quantum definition of locality. A subset of the observables within the algebra of a given causal diamond, corresponds to a smaller causal diamond included in the original one, if it is a tensor factor in the full algebra. The complementary factor is associated with the region of the large causal diamond outside of the smaller one. Note, that in contrast to quantum field theory, there are no mathematical subtleties about the tensor factorization, because everything is finite dimensional.
- The fact that the Hilbert space of a local region is finite dimensional has profound consequences. "Machines" that measure local physics are finite and have irreducible quantum fluctuations. As a consequence, precise answers to physical questions about a local region are not possible, *in principle*. This means that the mathematical description of this region (which e.g. will have some precise time dependent Hamiltonian) can never be completely verified and so has some degree of ambiguity. There are many equivalent descriptions of the same physics, compatible with our idea that in quantum gravity, local physics

[4] More precisely, we should discuss foliations of the null boundary by space-like $d-2$ surfaces. The claim is that there is a unique maximal area surface which can be chosen as a leaf of a foliation.

[5] Actually, observables are only the hermitian elements of the algebra. We will continue to use the phrase algebra of observables, with this implicit understanding.

[6] And thus has a unique unitary equivalence class of representations: the algebra of $N \times N$ matrices in a finite dimensional Hilbert space.

is *gauge dependent*. There may be some sense then, in which the holographic principle and the covariant entropy bound are the quantum origin of general covariance.

On the basis of these observations, W. Fischler and I attempted to formulate a general theory of holographic quantum space-time. The theory has three modules, two kinematic and the third dynamical. The two kinematical modules are called *the arena* and *the variables* and the dynamical module is called *the Hamiltonian constraint*.

1.1 The Arena

The arena consists of a collection of Hilbert spaces $\mathcal{H}_i$, which, in the limit of large dimension are to be thought of as the Hilbert spaces of causal diamonds whose holoscreen area is four times the logarithm of dim $\mathcal{H}_i$. For every pair of Hilbert spaces there is a tensor decomposition

$$\mathcal{H}_i = \mathcal{O}_{ij} \otimes \mathcal{N}_{ij},$$
$$\mathcal{H}_j = \mathcal{O}_{ji} \otimes \mathcal{N}_{ji},$$

where $\mathcal{O}_{ij} = \mathcal{O}_{ji}$. $\mathcal{O}_{ij}$ represents, in the limit where geometrical language is appropriate, the Hilbert space of the largest causal diamond in the geometrical overlap of the diamonds corresponding to the individual spaces. We insist that each $\mathcal{O}_{ij}$ be on the list of $\mathcal{H}_i$ and that the overlap of $\mathcal{O}_{ij}$ with either $\mathcal{H}_i$ or $\mathcal{H}_j$ is just $\mathcal{O}_{ij}$ itself. $\mathcal{N}_{ij}$ is the Hilbert space mapped out by all operators in $\mathcal{H}_i$, which commute with all operators in the overlap space. We also impose the condition that the overlap of $\mathcal{O}_{ij}$ with $\mathcal{H}_k$ be a Hilbert space $\mathcal{O}_{ijk}$ symmetric under permutations of i, j and k. We call this *the commutative rule for overlaps*. We should go on and define higher order multiple overlaps, obeying the generalized commutative rule. The process will stop after a while, because all of our Hilbert spaces (in a finite region of space-time) are finite dimensional.

Geometrically one should think of the arena as defining the quantum precursor of a covering of the Lorentzian space-time manifold by causal diamonds. If the covering is rich enough, one would expect it to be constrained, and to completely encode the space-time geometry. A Lorentzian geometry is determined up to a conformal factor by its causal structure, and a rich enough covering by causal diamonds would determine the causal structure. Furthermore, our association of Hilbert space dimensions with areas of holographic screens, obviously determines the conformal factor if the covering is rich enough.

The big surprise for those who thought that quantum gravity had some relation to a Feynman path integral over Lorentzian geometries is that the kinematically defined arena, already determines the space-time geometry. However, in order to ensure that the arena actually corresponds to some kind of space-time geometry we (must?) introduce two more postulates, relating to the existence of "observers".[7] Our current understanding of an observer in quantum mechanics is that it represents a large quantum system, approximately obeying the rules of cutoff quantum field theory. The observing devices available to an observer are the averages of local fields over volumes[8] large compared to the cutoff scale. These "pointer" variables have exponentially small quantum fluctuations. Obviously, any such observer will follow a time-like trajectory through space-time.

We define an abstract notion of observer in holographic quantum space-time by mimicking the last observation. In describing the variables of quantum gravity in the next subsection, we will introduce *the Hilbert space*, $\mathcal{P}$, *of a single pixel on a holographic screen*. We will define an observer to be a sub-collection of the $\mathcal{H}_i$ labeled by a positive integer n ranging up to some value n_{max}, which might be infinite. We insist that

$$\mathcal{H}(n) = \bigotimes \mathcal{P}^n .$$

Geometrically, we are constructing the analog of a nested sequence of causal diamonds, corresponding to larger and larger intervals on the observer's time-like world line.

The nature of the mapping between observer Hilbert spaces and causal diamonds in space-time will depend on the nature of the space-time. Two broad classes of space-times that we might want to find quantum descriptions of are Big Bang, and TCP symmetric, space-times. In the former, the Hilbert space $\mathcal{H}(n)$ will represent a sequence of causal diamonds whose past tips all lie on the Big Bang hyper-surface. As n increases, the future tip is a larger and larger positive time-like distance from the Big Bang. The space-time picture is shown in Fig. 1. The detailed nature of the mapping between n and some conventional geometrical measure of time evolution, will depend on the dynamical part of the theory, whose description we give later. One general point can be made. In a region of space-time which is roughly an expanding $k = 0$ FRW, the area of the holographic screen of a causal diamond grows like t^{d-2}, where t is the FRW time. The apparent discretization of time implied by our formalism is not fixed at the Planck length, but scales to zero as time goes on. Roughly speaking the minimal time interval is related by the time energy uncertainty relation, to the energy of the maximal black hole that could exist in the causal diamond at time t.

---

[7]Despite the traditional bio-centric language, our observers need not be living creatures. The only requirement is that they be quantum systems with a large number of observables that have very small quantum fluctuations. Observers have neither gender nor consciousness, only approximate local fields.

[8]Implicit in the idea that the observer is approximately described by field theory is the *assumption* that the region of space-time in which the observer lives is approximately classical. Later on, we will use the formalism we construct to describe situations in which no such semi-classical observer could exist.

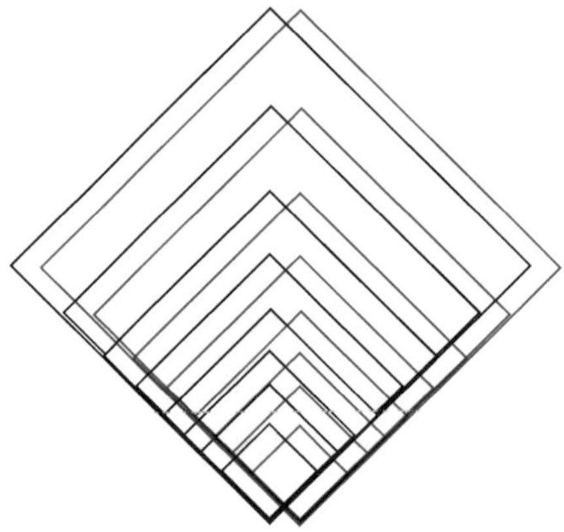

Fig. 1 Nested causal diamonds of a pair of observers in Big Bang space-time.

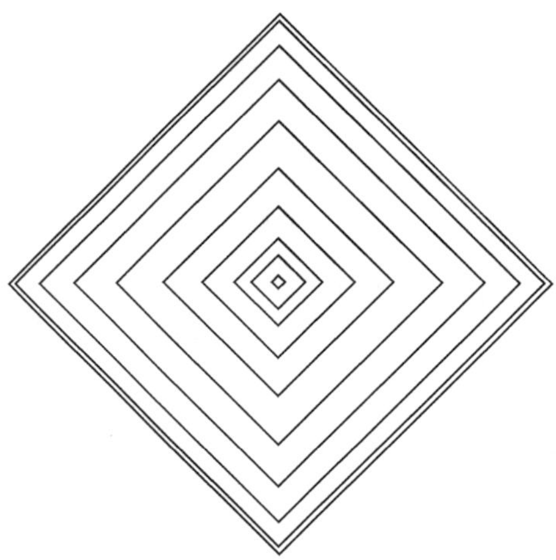

Fig. 2 Nested causal diamonds of an observer in TCP symmetric space-time.

By contrast, for TCP symmetric space-times we should view the sequence of Hilbert spaces $\mathcal{H}(n)$ as describing a nested set of diamonds centered around the point of time symmetry of the observer's trajectory. The picture is that of Fig. 2.

As we will see, in this context, the appropriate dynamical object is an approximate S-matrix, which relates in-coming to out-going states through the diamond's holographic screen. If $n_{max} = \infty$, we can hope to define a true gauge invariant scattering matrix.

Although asymptotically anti-deSitter space-times fit in this TCP invariant category, they introduce a new wrinkle. In space-times with a conformal diagram similar to that of Minkowski space, n goes to infinity with the time-like separation between the past and future tips of the diamond. It is well known however, that in AdS space, the area of the causal diamond of a geodesic observer, goes to infinity

Fig. 3 Causal diamonds in anti-deSitter space-time: the proper time/area relation becomes singular.

when the global time associated with this observer's trajectory reaches a single AdS period (Fig. 3).

After this time $n = \infty$ but the time continues to grow.[9] Thus, we expect a time-like evolution of a system with an infinite number of degrees of freedom. As noted above, the Bekenstein–Hawking entropy formula suggests, and years of work on the AdS/CFT correspondence confirm, that this system is a quantum field theory, living on the conformal boundary of AdS space.[10]

The alert reader will have noticed that TCP invariance is, in our formalism, not a general property of the laws of physics, but an *a priori* restriction on the class of space-times. This is entirely analogous to the restrictions imposed by insisting on Poincare or Anti-de-Sitter invariance. Symmetries in general relativity are *all* related to actions on a restricted set of boundary conditions. There is no reason to think that the quantum theory of a Big Bang space-time is related by symmetry to that of a Big Crunch. In the next section we will define a mathematical model of a

[9]In the universal cover of AdS, which has no closed time-like curves.

[10]Parenthetically, the analysis above may shed some light on the difficulties of extracting local physics from AdS/CFT. Bulk physics on scales much smaller than the AdS radius should be described by an approximate S-matrix. The connection of boundary time evolution with causal diamonds larger than the AdS radius suggests that the relation between this S-matrix and the boundary Hamiltonian might be complicated. It would be interesting to revisit the proposals of Polchinski [12] and Susskind [13] for extracting the flat space S-matrix from CFT correlators, in view of this insight.

Big Bang space-time, which is definitely not related to a Big Crunch. Despite the intrinsic violation of TCP in the quantum theory of gravity, we will see that we do not automatically have an explanation of the thermodynamic Arrow of Time (AOT).

One observer does not a space-time make. We will model a non-compact space-time, which admits a foliation by non-compact space-like hypersurfaces. To this end, we introduce a $d - 1$ dimensional lattice, whose points are labeled by a variable $\mathbf{x}$. Only the topology of the lattice is relevant: our definition of distance will not be that of the lattice. We will, for the moment, take it to have the topology of Euclidean space. To each lattice point we attach an observer, so that we have a collection of Hilbert spaces $\mathcal{H}(n, \mathbf{x})$. We must now add to this collection, a collection of overlap Hilbert spaces $\mathcal{O}(n, m, \mathbf{x}, \mathbf{y})$, as well as multiple overlaps obeying the commutative rule. We postulate that $\mathcal{H}(n, \mathbf{x})$ has the same dimension for all lattice points $\mathbf{x}$. We are describing a space-time in terms of a congruence of time-like trajectories, and this rule simply defines a particular time slicing, which we call *equal area slicing*. We further postulate that if $\mathbf{x}$ and $\mathbf{x}_{nn}$ are nearest neighbor points on the lattice, that the dimension of $\mathcal{O}(n, n, \mathbf{x}, \mathbf{x}_{nn})$ is $\dim \mathcal{P}^{n-1}$. This is what it means for two observers' trajectories to be nearest neighbors, for all time. Their available information, at any time, differs only by one pixel's degrees of freedom. The specification of more complicated overlaps, as we will see, is enmeshed with the dynamical equations.

The reader, especially if she is a string theory aficionada, will be getting agitated at this point. This all looks like a rather dreary translation of simple-minded space time concepts into statements about a quantum theory. What about compact dimensions? What about Stringy Geometry and dualities? It is important to emphasize that our discussion so far is describing only the non-compact dimensions of space-time, where I include a single horizon volume of de Sitter space in the phrase non-compact. Extra, compact, dimensions will be encoded in the algebra of pixel variables, to which we now turn.

1.2 The Variables

The key notion in the holographic approach to quantum gravity is the connection between the dimension of the Hilbert space of a causal diamond, and the area of its holographic screen. It is natural to think of the structure of the holographic screen as defining the set of geometrical degrees of freedom which appear in the quantum theory of gravity. So we must first ask what kind of space the holographic screen is. Mathematicians have long known that the topology and measurability properties of a space can be encoded in its algebra of functions. In classical topology, the topological properties are encoded in the C^* algebra of continuous complex valued functions, while the measure properties of the space are encoded in its von Neumann algebra of measurable functions. The basic idea of non-commutative geometry is to consider "spaces" whose algebra of functions is a non-commutative C^* or von Neumann algebra.

Geometrically, a *pixelation* of a smooth space, replaces it by a finite set of exactly flat regions, which form some approximation to the original space. This can be made precise by stating it in terms of the algebra of functions. A basis of functions on a pixelated space consists of the functions which are equal to 1 on a single pixel, and vanish on all the others. So, in terms of function algebras, a pixelation is the replacement of the algebra of continuous or measurable functions by a finite dimensional algebra. Naive geometrical pixelations always correspond to a commutative algebra, isomorphic to an algebra generated by commuting orthogonal projectors on a finite dimensional vector space. We will consider more general, non-commutative, finite dimensional approximations to the function algebras of holographic screens,such as fuzzy spheres [14].

Think about a small, geometrical pixel on a holographic screen. Up to a conformal factor, the geometrical information in this pixel consists of a pair of null vectors, one ingoing and one outgoing (through the holoscreen surface) and a little piece of transverse hyperplane. Precisely this information is encoded in a spinor satisfying the Cartan–Penrose equation:

$$\bar{\psi}\gamma_\mu\psi(\gamma^\mu)_{ab}\psi_b = 0.$$

This equation implies that the vector $\bar{\psi}\gamma_\mu\psi \equiv n_\mu$, is null, and that the spinor ψ is a transverse spinor for this null vector. The tensors $\bar{\psi}\gamma_{(\mu_1\ldots\mu_n)}\psi$, all lie in a $d-2$ dimensional transverse hyperplane, and define the orientation of that hyperplane.

The CP equation is Lorentz invariant and invariant under the rescaling $\psi \rightarrow \lambda\psi$.[11] We will treat both of these as gauge equivalences, rather than global symmetries, in accordance with the idea that global symmetries in general relativity are properties of specific kinds of asymptotic boundary conditions, and not of the general local formalism. These two gauge equivalence groups have very different roles to play and we will treat them quite differently. We will pick a physical gauge for local Lorentz invariance, reducing the number of degrees of freedom on each pixel n to the number, in a transverse spinor, $S_a(n)$.[12] The Lorentz connection relating spinor bases on different pixels will be encoded in the Hamiltonian we write for the physical degrees of freedom. This is analogous to the H connection in the gauge invariant formulation of a G/H sigma model, which is just a function of the sigma model degrees of freedom.

By contrast, the rescaling invariance of the CP equation will be explicitly broken to a Z_2 subgroup by our quantization procedure. Indeed, this redundancy arises because the CP equation only expresses the orientation of the null vector n_μ and of its transverse hyperplane. In geometrical language, it is insensitive to the conformal factor of the metric. Rescaling invariance is a direct expression of this fact. The

---

[11] λ is real or complex, depending on the character of the minimal spinor representation in dimension d.

[12] Recall that for a general pixelation of the function algebra, the label n stands for a single element in the basis of the finite dimensional algebra.

remaining geometrical information arises from the Bekenstein–Hawking connection between entropy and area, which is a quantum relation. Thus we may expect that the quantum commutation relations between the $S_a(n)$, break the scaling symmetry of the CP equation. If we want to preserve the residual $SO(d-2)$ symmetry of the fixed Lorentz gauge, and have a finite number of states for a pixel, the unique quantization rule is

$$[S_a(n), S_b(n)]_+ = \delta_{ab}.$$

This formula is appropriate in dimensions where the spinor of $SO(d-2)$ is real. When it is complex, we instead have the commutators of creation and annihilation operators. Note that these anti-commutation relations break the scaling symmetry of the CP equation to a $Z_2 : S_a(n) \to -S_a(n)$. The operators associated with independent pixels should commute with each other, consistent with the idea that quantum gravity is a system with independent degrees of freedom per unit area of the holographic screen. However, we can use the Z_2 gauge invariance to perform a Klein transformation, and obtain pixel operators satisfying

$$[S_a(n), S_b(m)]_+ = \delta_{ab}\delta_{mn}.$$

There is an obvious linear action of the pixelated algebra of functions on the set of operators $S_a(n)$. Thus, if f_n is a basis of the function algebra we define

$$f_n S_a(k) = C^{l}_{nk} S_a(l),$$

where the structure constants are defined by

$$f_n f_k = C^{l}_{nk} f_l.$$

Thus, in the limit where the dimension of the function algebra goes to infinity, and approaches the algebra of functions[13] on a smooth $d-2$ manifold (the classical holographic screen of a large causal diamond) the limit of the $S_a(n)$ will be some sort of operator valued section of the spinor bundle over the holographic screen.

1.3 SUSY and the Holographic Screens

The most remarkable thing about this general holographic approach to quantum gravity, is that we are close to showing that a theory of quantum gravity, defined according to our rules, automatically incorporates Supersymmetry (SUSY). To

[13] We do not yet specify the properties of these functions, with regard to continuity, but they should at least be measurable.

make this more precise, let us specialize to the case of asymptotically flat space-times. In this case, the holoscreens of causal diamonds of geodesic observers are round $d-2$ spheres. Imagine that the limiting $S_a(n)$ operators should be thought of as linear functionals (measures) on the space of measurable sections of the spinor bundle. A basis thus has the form

$$S_a(n) \to S_a \delta^{d-2}(\mathbf{\Omega}, \mathbf{\Omega}_0),$$

where $\mathbf{\Omega}_0$ is a point on the sphere, and

$$[S_a, S_b]_+ = \delta_{ab}.$$

Thus, the degrees of freedom associated with a point $\mathbf{\Omega}_0$ are those of a massless superparticle with momentum $p(1, \pm\mathbf{\Omega}_0)$. Note that nothing we have said so far specifies the value of p.

This exciting observation poses a number of immediate questions

- It is not sensible that the quantum theory of gravity is just a theory of one particle. Where are multi-particle states?
- What guarantees that the super-multiplet contains the graviton?
- What about massive particles?
- Where is the variable describing the overall scale $p \geq 0$ of the massless particle momentum?
- What is the meaning of the $\pm$ ambiguity in the association of massless momenta with the parameter $\mathbf{\Omega}_0$ labeling pixel operators?

The penultimate question, and the first one, receive a common answer via the following ansatz: Rather than allowing the the algebra of a finite causal diamond to converge to the algebra of measurable functions on the sphere, we instead require that it converge to the tensor product of this algebra with the Cartan subalgebra[14] of the unique hyperfinite II_∞ factor, $R[0, 1]$, of Murray and von Neumann [15]. This algebra is generated by finite sums of commuting orthogonal projection operators e_i. It has a trace function which takes on all values in $[0, \infty]$. Two projectors with the same trace are unitarily equivalent via a unitary operator in $R[0, 1]$, and the permutation of a collection of projectors with the same trace is also a unitary. The group of outer automorphisms of the algebra is a one parameter group, which rescales the trace

$$\mathrm{Tr}\, \rho_t(e_I) = e^t \mathrm{Tr}\, e_I.$$

[14] I am not sure whether this terminology is standard for the maximal abelian subalgebra of an associative, rather than a Lie, algebra. Instead of the Cartan subalgebra, we could use the whole non-abelian von-Neumann algebra, but insist that inner automorphisms of the algebra are gauge transformations.

The inner automorphisms (unitary transformations) of $R[0,1]$ are considered to be gauge equivalences of the physical theory.[15]

The quantum operator algebra consists of all finite collections of operators

$$S_a(p_I)\delta(\mathbf{\Omega}, \mathbf{\Omega}_I),$$

with

$$[S_a(p_I), S_a(p_J)]_+ = \delta_{ab}\mathrm{Tr} e_I e_J.$$

The permutation gauge symmetry is to be implemented as a constraint on quantum states. This is similar to the operator algebra that describes the asymptotic states of Matrix theory (if $d = 11$). Note however that each asymptotic massless particle is treated in the light cone frame along its momentum direction, so there are no transverse momenta in the specification of the Hilbert space. As in matrix theory, the permutation symmetry of particle statistics is the gauge equivalence between diagonal matrices mod permutations, and commuting matrices. The II_∞ factor is the appropriate infinite limit of the space of matrices.

If we take the limit of the operator algebra of a finite causal diamond in this way, we get an algebra whose smallest representation is the Fock space of a collection of massless superparticles, with the correct spin-statistics connection. The spin states of superparticles penetrating the holographic screen are, in this interpretation, just the quantized degrees of freedom of the screen itself.

There is an ambiguity, a quite standard one, in the representation of this operator algebra. For single particle states, with fixed momentum, the operator algebra is isomorphic to that of a finite number of fermionic creation and annihilation operators. The "ground state" of the algebra can be chosen to lie in any representation of the group $SO(d-2)$ (there are sometimes additional constraints from the requirement that the spectrum is TCP invariant). In supersymmetric four dimensional particle physics, this ambiguity allows us to discuss massless supermultiplets with any maximum helicity. However, well known theorems for asymptotically flat space [16] show that we cannot have a non-trivial scattering matrix for particles which have four dimensional helicity greater than two. On the other hand, since we are trying to construct a theory of gravitation, we should expect to have gravitons in the spectrum.

For $d = 11$, there are 16 real generators and the minimal representation of the algebra consists of precisely the graviton supermultiplet. The construction we have outlined leads precisely to the Fock space of 11D SUGRA as the Hilbert space of an infinite causal diamond [17]. I believe that the correct way to obtain more general theories of quantum gravity in asymptotically flat space, is to view them as compactifications of this example. This will guarantee that there is always a graviton in the spectrum, and no higher spin massless particles.

[15] As noted above we could in fact have required our algebra to be the tensor product of the full II_∞ factor with the algebra of functions on the sphere, subject to this gauge equivalence.

As an example, consider compactification on a k dimensional torus. The 16 pixel operators are denoted S_a^I, where a is a spinor index of $SO(9-k)$. The algebra of functions on the k torus is pixelated by replacing it by the algebra of functions on $Z_{N_1} \times \ldots \times Z_{N_k}$, and a basis is labeled by a k vector $(P_1 \ldots P_k)$, whose jth component is in Z_{N_j}. If n labels pixels in the non-compact holographic screen (which approaches S^{9-k} for large causal diamonds) then

$$[S_a^I(m), S_b^J(n)]_+ = \delta_{ab}\delta_{mn}(\delta^{IJ} + (\vec{\gamma})^{IJ} E \vec{P},$$

where $\vec{\gamma}$ are the k dimensional Euclidean Dirac matrices and E is the k-bein for the torus. If $k \geq 2$ this algebra should be modified to include wrapped brane charges. Note that, with this prescription, the number of degrees of freedom per non-compact pixel, depends on the spectrum of the operators P_j which label a basis of the function algebra on the compact space. This should be viewed as the renormalization of the Planck scale upon compactification:

$$\left(m_P^{(11-k)}\right)^{9-k} = V_k \left(m_P^{11}\right)^9 .$$

I have not worked out the limiting process, which gives the Fock space of BPS particle states of M-theory compactified on a k torus, but this should be relatively straightforward. More interesting would be the construction of theories with less than maximal SUSY. Here, the "idea for an idea" is to keep the number of operators per pixel of the non-compact holographic screen equal to 16, in order to guarantee the existence of gravitons, but to allow operators on the right hand side of the anti-commutation relations, which do not commute with the S_a^I. Thus, the properties of the compact space will be encoded in the pixel algebra. Notice that in current descriptions of M-theory compactifications, it is those aspects of compactification that appear in the SUSY algebra, which are invariant under dualities.[16]

These examples already introduce the question of massive particles. In the limit of large causal diamonds, we will obtain single particle operators $S_a^I \delta(\mathbf{\Omega}, \mathbf{\Omega}_0)$, where $\mathbf{\Omega}$ is a point on the $9-k$ sphere. From the incoming and outgoing positive energy null momenta pointing in this direction, we can construct $p(1, \mathbf{\Omega}) + \alpha \mathbf{p}(1, -\mathbf{\Omega})$, with positive α. This is a massive four vector, with mass

$$4p^2\alpha.$$

Thus the relation between the positive parameter p (determined by the trace of the projector in the II_∞ algebra and the spatial momentum in the $\mathbf{\Omega}$ direction is

$$\mathbf{P} = \left(p - \frac{m^2}{4p}\right)\mathbf{\Omega}.$$

[16]This is not precise. The torsion elements of K-theory, which classify stable, non-BPS particles, do not appear in the SUSY algebra.

For massive particles, small p corresponds to negative momentum. In [17] I showed that Lorentz boosts along the direction $\mathbf{\Omega}$ were implemented with the help of the outer automorphism of the II_{∞} factor which re-scales p. For a massive particle, a Lorentz boost along its direction of motion can reverse the sense of the particle's motion, so the fact that rescaling p changes the sign of the momentum is expected.

The parameter m^2 which appears in the previous paragraph, will enter the formalism through the definition of the conserved SUSY generators (and thus the non-compact energy-momentum vector). For BPS states, its value is determined algebraically. However, string theory also contains stable particles with torsion class K-theory charges, whose mass is not determined by group theory. The kinematical formalism we are discussing must take these masses as input parameters, which are to be determined by the dynamical equations.

Recall that we admitted a sign ambiguity in the assignment of a null momentum to a pixel on the holoscreen of a large causal diamond. This ambiguity is the distinction between ingoing and outgoing particles. Thus, we should imagine two copies of the kinematical algebra of pixel operators, related by the scattering matrix. A key part of the theory, which has not yet been constructed, is a set of dynamical equations for the scattering matrix.

1.4 The Hamiltonian Constraint

Classical general relativity is invariant under general coordinate transformations. In particular, the transformations that change between different time slicings of the space-time manifold, including moving to the "next" slice of a given slicing, are gauge transformations. This is expressed through the Hamiltonian constraint equations

$$H(x) = 0,$$

$$[H(x), H(y)]_{PB} = (P^i(x) + P^i(y))\partial_i \delta(x, y).$$

x, y are coordinates on a given time slice and $P^i(x)$ generates diffeomorphisms of the slice. $H(x)$ generates a small deviation into the future, at the point labeled by x. There are many ways to push a particular point a finite distance into the future. The Poisson bracket relation is a consistency condition which ensures that, independently of how one moves to the future, the diffeomorphism invariant quantities at a given space-time point are the same.

In the most naive approach to the quantization of gravity as a field theory, the Hamiltonian constraint is imposed as an equation for physical states

$$H(x)|\psi\rangle = 0.$$

When gravity is treated semi-classically, this leads to quantum field theory in curved space-time, with a (generally time dependent) Hamiltonian generating evolution along some time parameter of the classical background space-time [18, 19].[17]

These semi-classical ideas suggest a dynamical principle for the arena. I will state it for cosmological space-times, which begin in a Big Bang. Here one takes the nested causal diamonds describing a single observer to all begin at the same point. The system has a time dependent Hamiltonian, which evolves things from the Big Bang onward. The Hamiltonians $H(n)$ operate in the full Hilbert space of the maximal causal diamond of the observer. $e^{-iH(n)}$ generates time evolution between the tip of the $n-1$st causal diamond and the nth. The time evolution is discrete, but the implicit time interval becomes shorter and shorter as n, and the dimension of $\mathcal{H}(n, \mathbf{x})$ get larger. For example, in an FRW universe with spatially flat time slices, the area/entropy, which increases by $\ln \dim \mathcal{P}$ when n goes to $n+1$, grows as t^{d-2} in FRW time. Roughly speaking, the discreteness in time is inversely related to the energy of the maximal size black hole that can be encompassed by the causal diamond, by something like the time-energy uncertainty relation.

We insist that the Hamiltonian $H(n)$ be constructed as a sum of operators $H_{in}(n) + H_{out}(n)$. $H_{in}(n)$ contains only pixel operators $S_a(n, \mathbf{x})$ for the nth causal diamond, while $H_{out}(n)$ is constructed out of operators which commute with all of those pixel operators. This postulate imposes causality on the time evolution. If we view the full evolution on the Hilbert space of the maximal causal diamond, then at the time labeled by n, degrees of freedom inside the causal diamond have not yet interacted with those outside it. Note that this framework has both a natural beginning of time (where the number of accessible degrees of freedom shrinks to zero) and a natural Arrow of Time.

The natural Arrow of Time is in fact already a feature of the Wheeler–DeWitt formalism. We build a Hilbert space for the WD equation by expanding around a particular classical solution, which is usually not time reversal invariant. Adherents of the fallacious point of view that general relativity is just another quantum field theory will try to argue that this is some kind of *spontaneous* breaking of TCP invariance [19], since Einstein's Lagrangian is locally invariant. The remarks of the introduction should prove an adequate refutation of this mistaken point of view. Different solutions of the same low energy effective gravitational action *are not* part of the same quantum theory, unless they correspond to small perturbations of a given macroscopic geometry. The conclusion is that, like all real symmetries in GR and QG, TCP invariance is a *global*/asymptotic symmetry. If it is not a property of the macroscopic geometry, it is not a property of the theory.[18]

---

[17] An often overlooked part of this analysis is that the Hamiltonians and Hilbert spaces for different classical solutions are not related to each other in any simple way.

[18] It is perhaps worthwhile to point out a linguistic nicety. We have become used to using the word *theory* to mean a particular mathematical model, defined by a fixed evolution operator $U(t, t_0)$, and I am using the term in this sense. The general THEORY of quantum gravity is the broader subject of this review, which is illustrated by a variety of particular theories. It's clear that in a

Now consider the sequence of overlaps $\mathcal{O}(n, \mathbf{x}; n\mathbf{y})$. For consistency with the geometrical picture (Fig. 1) each successive overlap should contain the previous Hilbert space as a tensor factor. Even if we assume pure state initial conditions for the Hilbert space of each observer we should not expect the states in the overlap Hilbert spaces to be pure. The Hamiltonians $H(k, \mathbf{x})$ for $k \leq n$ are not sums of operators which act only in $\mathcal{O}(n, \mathbf{x}; n, \mathbf{y})$ and operators which act only in its tensor complement in $\mathcal{H}(n, \mathbf{x})$. Instead, the overlap is, from the point of view of the observer at $\mathbf{x}$, described by a sequence of density matrices, $\rho(n, \mathbf{x}; \mathbf{y})$ constructed by evolving the pure state in $\mathcal{H}(n, \mathbf{x})$ and tracing over the tensor complement of $\mathcal{O}(n, \mathbf{x}; n\mathbf{y})$. Similarly, we can use the evolution in $\mathcal{H}(n, \mathbf{y})$ to construct $\rho(n, \mathbf{y}; \mathbf{x})$. The fundamental dynamical consistency condition is that there exist unitary operators such that

$$U^{\dagger}(n, \mathbf{x}; \mathbf{y})\rho(n, \mathbf{x}; \mathbf{y})U(n, \mathbf{x}; \mathbf{y}) = \rho(n, \mathbf{y}; \mathbf{x}),$$

for every $n, \mathbf{x}, \mathbf{y}$. This is an infinite set of staggeringly complicated conditions.

In my opinion it is plausible that any solution of these conditions defines a consistent model of quantum cosmology. Similarly, if we consider TCP symmetric situations, each observer is associated with an infinite sequence of "partial S matrices" relating the incoming and outgoing states on successive causal diamonds, which converge to the gauge invariant S-matrix.[19] There will be analogous consistency conditions for the overlap Hilbert spaces and these partial S-matrices. The interesting difference is that EACH observer's sequence of S-matrices should converge to the same answer. The consistency conditions should lead to some sort of equations for the gauge invariant S-matrix.

One of the most interesting ideas that came out of the ancient era of S-matrix theory was that Poincare invariance, crossing symmetry, unitarity, and a vaguely defined notion of analyticity, would determine the S-matrix uniquely and possibly even compute the spectrum of particle masses. We now know that this is wrong. Indeed, S matrix theorists proved in the 1960s, that in perturbation theory, with a finite number of stable particles, these conditions led exactly to the ambiguity of local field theory. We now know that there are many non-perturbative solutions of these constraints, defined by relevant perturbations of conformal field theories, which flow to massive fixed points. However our holographic theory of quantum gravity suggests stronger constraints on the S-matrix. First of all, we expect to have a graviton in the spectrum. Secondly, we expect exact super-Poincare symmetry, though the supergroup is not precisely specified when the dimension of space-

---

sensible world we would use the word model instead of theory in phrases like "the ϕ^4 quantum field theory".

[19]In AdS space-time this statement is modified, because the label n representing the area of the causal diamond, becomes infinite for finite proper time separation between tips of the diamond. It is not clear what the infinite sequence of partial S-matrices converges to. The familiar boundary dynamics of these space-times most naturally describes what happens to causal diamonds after they hit the time-like boundary and have infinite area.

time allows for an ambiguity. It does not seem implausible that the constraints on the S-matrix that follow from our overlap conditions, combined with symmetries and unitarity, will determine the scattering matrix up to the sets of parameters we already know about. For models with at least 16 supercharges, everything will be parametrized by the known moduli spaces. For 8 supercharges (M-theory compactified on a Calabi–Yau three-fold), we still do not have a global picture of the moduli spaces.

The challenge of finding equations for the S-matrix, which complete the old S-matrix program and reproduce the results we know from string dualities is the first step in turning the kinematic framework advocated here into a real theory of quantum gravity. Armed with the solution of that problem we would be much further along the road to finding a complete quantum theory of the real world.

Since I have not completed that program, I can record here only the progress that has been made in the kinematic framework. In the next section I will describe the dense black hole fluid (DBHF) cosmology, which is the only known mathematically complete solution to the consistency conditions for holographic space-time. Its coarse grained description is that of a flat FRW cosmology, with equation of state $p = \rho$. I then outline an heuristic derivation of normal cosmology from a description of defects in the DBHF. One of the results of this is the prediction that the future asymptotic state of the universe is a de Sitter (dS) space, with cosmological constant (c.c.) determined by the number of degrees of freedom associated with the initial defect. If one considers the mathematical description of the universe to consist of an infinite DBHF, sprinkled with a collection of defects of various sizes, holographic cosmology leads to a multiverse, where different normal regions of the universe have different positive values of the c.c. This gives us the possibility of using environmental/anthropic constraints to explain the peculiar value of the c.c. that we observe. The extent to which other parameters are environmental accidents, depends on the uniqueness, in the limit of small c.c., of the quantum theory of dS space. Since the model with vanishing c.c. must be a super Poincare invariant system with compact moduli space [20], and no such models are known as yet, it may be that the limit is quite unique. This could lead to a predictive theory of particle physics and cosmology, with only the c.c. determined by anthropic constraints. Further, we will see below that this framework may put much stronger anthropic constraints on the c.c. because many other particle physics parameters are functions of it.

1.5 Executive Summary

The quantum theory of gravity is not a quantum field theory because its typical localized excitation is a black hole, which satisfies a non-field theoretic relation between entropy and geometry. We proposed instead a theory where the basic operator algebra associated with a causal diamond was constructed from variables describing the orientation of pixels on the holographic screen of the diamond. For large diamonds the variables associated with a pixel describe the spin information in a supersymmetry multiplet and are to be associated with particles entering or leaving the holoscreen via that pixel. The geometry of finite area holoscreens is determined by a finite dimensional associative algebra and a pixel is a particular basis element of this algebra. The pixel variables are operator valued sections of the spinor bundle over this algebra.

Space-time causal relations are encoded in the tensor factor structure of the Hilbert spaces associated with different diamonds. The overlap region between two diamonds is associated with a common tensor factor in the Hilbert spaces of the individual diamonds. This leads to complicated consistency conditions on the dynamics of the system. The conformal factor of the space-time is determined in terms of Hilbert space dimensions, by the Bekenstein–Hawking rule. We defined a quantum space-time in terms of the causal diamonds associated with the trajectories of a set of densely spaced time-like observers, with dynamics obeying the consistency conditions.

Easier said than done. In the next section I will present the only known solution to the consistency conditions, an idealized cosmology called the dense black hole fluid (DBHF). This gives a derivation of a particular Friedmann–Robertson–Walker geometry from a system with a purely quantum mechanical definition. I then discuss an heuristic model of real cosmology, whose initial state may be described as a defect in the DBHF. I show how homogeneity, isotropy, flatness, and low entropy in localized degrees of freedom, follow naturally from this model. Furthermore, it predicts that the universe will asymptote to a de Sitter state, with c.c. determined by the initial conditions. I discuss the possible origin of microwave background fluctuations in this model, as well as the origin of the Arrow of Time.

2 Holographic Cosmology

2.1 *The Dense Black Hole Fluid*

The idea for the DBHF cosmology began with the observation of Fischler and Susskind [21] that a flat FRW cosmology with $p = \rho$ could saturate the covariant entropy bound at all times. The question that arises is what this peculiar substance is. Let me begin by discarding a red herring. Consider a general sigma model, with Lagrangian

$$G_{ij}(\phi) g^{\mu\nu} \sqrt{-g} \partial_\mu \phi^i \partial_\nu \phi^j \,.$$

An FRW model based on this Lagrangian, coupled to Einstein gravity, has scalar fields which depend only on FRW time. Since the Lagrangian has no potential, and no spatial gradients are excited, the equation of state is $p = \rho$. However, this cosmology has zero entropy density. The entire universe has only a finite number of degrees of freedom excited, and they evolve along a fixed classical trajectory. If we consider generic low energy excitations of this model, we get a massless gas, with $p = \rho/(d-1)$, in d dimensional space-time.

An heuristic idea of what is involved in getting a substance that saturates the entropy bound comes from considering an instantaneous configuration of black holes, separated by a distance of order their Schwarzschild radius. The entropy density is (in Planck units)

$$\sigma \sim R_S^{-(d-1)} R_S^{d-2},$$

while the energy density is

$$\rho \sim R_S^{d-3} R_S^{-(d-1)} \sim \sigma^2.$$

For an extensive fluid, the first law of thermodynamics (equivalently, covariant conservation of the stress tensor in FRW cosmology) reads

$$\rho dV + V d\rho = T\sigma dV + VT d\sigma - p dV,$$

whence

$$\rho = T\sigma - p,$$

and

$$d\rho = (p + \rho)/\sigma d\sigma.$$

For $p = \rho$, this implies $\rho \propto \sqrt{\sigma}$, just like the dense black hole fluid. Thus, the DBHF is a system of black holes which coalesce as the universe expands, in such a way that $\rho = k\sqrt{\sigma}$ at all times. This system has an equation of state $p = \rho$.

To construct an actual mathematical model of this system we consult our general formalism, and begin by concentrating on a single observer's Hilbert space.[20] A way to ensure that that the model maximizes the entropy available inside the horizon is to supply it with a random time dependent Hamiltonian. We will do so, but we choose our random ensemble in a manner motivated by the pixel variables of quantum gravity.

Ignoring subtleties associated with compact dimensions, the pixel variables inside the horizon at step k have the algebra

$$[S_a(m), S_b(n)]_+ = \delta_{ab}\delta_{mn},$$

where a, b are indices in the spinor representation of $Spin(d-2)$, with d the number of non-compact space-time dimensions. m and n range from 1 to k and label a basis for the algebra of functions on the pixelated holographic screen. The

[20]Once the model is constructed, it will be clear that the resulting universe cannot actually have any observers in it. Nonetheless, it will be describable by the formalism we constructed to model the experience of a normal observer.

Hamiltonian $H(k_{max})$ is constructed from all of the pixel variables, with $1 \leq n \leq k_{max}$. If d_S is the dimension of the spinor representation, then $k_{max}\ln d_S$ is the maximal entropy accessible to this observer. $H(k)$ is the sum of an operator $H_{in}(k)$ involving only $1 \leq n \leq k$ and another, $H_{out}(k)$ involving only $k+1 \leq n \leq k_{max}$. For the DBHF it will turn out that we only need to specify H_{in}.

We begin by choosing the term in H_{in} quadratic in pixel operators to be

$$H_{in}^2(k) = \sum S_a(m)h(m,n;k)S_a(n).$$

For each k, we make an independent choice of $h(m,n;k)$ from the orthogonal ensemble of anti-symmetric $k \times k$ matrices. Note that in writing a scalar product between spinor indices associated with different pixels, we have implicitly assumed that there is a flat $Spin(d-2)$ connection on the holographic screen. In a moment, we will present evidence that in the large k limit, the emergent space-time we are constructing has homogeneous and isotropic spatial slices. The holoscreen is a round sphere, and $SO(d-1)$ rotations supply the required connection.

Wigner's semi-circle law for the orthogonal ensemble, implies that the spectrum of our Hamiltonian for large k is that of a cut-off free massless Majorana fermion in $1+1$ dimensions.

Polynomials of order 6 or higher in the pixel operators are irrelevant perturbations of the free fermion fixed point, as long as their coefficients are bounded by k independent constants. We generalize our ansatz by adding an arbitrary, k dependent irrelevant perturbation to the Hamiltonian. The result of this construction is that, for large k, independent of the initial state, the system explores the full Hilbert space of the causal diamond at step k, during its evolution from the Big Bang. All degrees of freedom inside the horizon are equilibrated, and those outside are decoupled from them.

To make a fully consistent quantum cosmology, we have to add a full $d-1$ dimensional lattice of observers and definite overlaps and dynamics satisfying all the overlap conditions. Label the lattice points by $\vec{x}$. The equal area time slicing tells us that

$$\mathcal{H}(k, \vec{x}) = \bigotimes \mathcal{P}^k,$$

where $\mathcal{P}$ is the irreducible representation of the single pixel algebra. We choose the Hamiltonians

$$H(k, \vec{x}) = H(k),$$

where $H(k)$ is *the same* random Hamiltonian for all $\vec{x}$.

Our overlap Hilbert spaces at equal time are defined to be

$$(k, \vec{x}, \vec{y}) = \mathcal{H}(k - d(\vec{x}, \vec{y})),$$

where $d(\vec{x}, \vec{y})$ is the minimal number of lattice steps between the two points. Note that for large enough d the argument becomes negative. We interpret this to mean that the overlap is empty. That is, for fixed k only some of the points on the lattice are in causal contact (note that the latter phrase makes sense in our quantum system without any reference to geometry).

If we choose the same initial density matrix for each observer then this system satisfies the infinite set of overlap conditions we discussed above. Thus, despite the fact that the initial state of all degrees of freedom that are ever observable by any observer is freely chosen, we find a homogeneous cosmology. All observers experience the same sequence of states. We can inquire whether it would be possible to have an inhomogeneous cosmology, consistent with the random Hamiltonian ansatz for individual observers. This seems very implausible: the infinite set of overlap conditions are not satisfied if one allows the Hamiltonians of different observers to be different. The claim then is that the DBHF is automatically a homogeneous cosmology, if it is a cosmology at all.

To prove that it is, we *define* the coordinate distance between two points on our lattice, at fixed time, to be $d(x, y)$. By "unfolding the carpenter's rulers" we see that as $k \to \infty$, the geometry that this defines becomes isotropic. Furthermore, since it has no scale apart from the k dependent "horizon radius" outside which overlaps vanish, there is no spatial curvature. Thus, in the large k limit our cosmology is a flat FRW model.

Further support for this contention comes from examining how the horizon radius scales with the entropy. This is conveniently done by introducing *cosmological time*. If our model is indeed related to a flat FRW model, then the area of the horizon scales as t^{d-2}. On the other hand, the horizon entropy in our model scales like k, the number of copies of the pixel algebra. We conclude that we should define $t \propto k^{\frac{1}{d-2}}$. The physical volume of the horizon then scales like $k^{d-3/d-2}$, according to the geometrical model.

Our model contains a quantum mechanical definition of the coordinate distance to the horizon. It is the minimum number of lattice steps to the point where the overlap Hilbert space is empty. This is $s = k$. There is also an intrinsic definition of the physical distance traversed in one lattice step at time k, which scales like t^{3-d}. Thus, the distance to the horizon scales like t, which is what we would derive from a flat FRW model.

2.2 *Our World As the Cosmology of DBHF Defects*

The model of the previous subsection is nice mathematics, but is obviously not a theory of the real world. Nonetheless, it has some aspects that make it an attractive candidate for the description of the beginning of the universe or Big Bang. First of all, it represents the most generic possible initial state of a theory of quantum gravity, if we accept the covariant entropy bound. Secondly, it is well known that the stiffest fluid available dominates the energy density in an FRW universe, and we have shown that for large causal diamonds the DBHF behaves like a flat FRW universe with the maximally stiff equation of state $p = \rho$. Thus, we can view the DBHF as a quantum resolution of the singular Big Bang cosmology with $p = \rho$ and flat spatial sections.

To obtain a model universe which could have real observers in it,[21] we try to modify the maximally generic DBHF in some initially minor fashion. We can then view our modification as a *defect* living in the space-time defined by the DBHF. From the mathematical point of view, what we would like to do is to modify the assignment of overlaps and Hamiltonians, for some finite set of points on our lattice, in a way that is compatible with all of the overlap consistency conditions, but allows regions of the lattice to behave (for large k) like a quantum field theory in a homogeneous state on some background classical space-time.

Unfortunately, we have not been able to solve this problem. Our proposal is based instead on geometric intuitions. We view the normal regions embedded in the DBHF as arising from smaller than horizon filling black holes, which decay into a normal, radiation dominated FRW universe. We then ask about the evolution of an inhomogeneous system, in which these normal regions are embedded in a dense black hole fluid. In accordance with our general setup, we work on time slices such that in each portion of the universe, the area of an observer's causal diamond, reaching back to the Big Bang, is the same.

Let us start with a spherical region of normal fluid, embedded in the DBHF. Its coordinate size is $L_1(t)$, and this must grow at least as fast as the coordinate distance to the horizon if physics in this region is to remain normal. Thus $L_1(t) > ct^{\frac{1+3w}{1+w}}$. w is the asymptotic future equation of state parameter for the normal region, and we begin with the assumption $-1 < w < 1$. On equal area time slices, the cosmological time in the normal region is related to the cosmological time in the DBHF region by

$$t = \gamma T,$$

$$\gamma = \frac{1+3w}{2(1+w)}.$$

The first Israel junction condition imposes continuity of the metric at the interface:

$$a(t)L_1(t) = (t/t_0)^{1/3}L(t),$$

where $L(t)$ is the coordinate size of the spherical hole in the DBHF. For the indicated range of w, $a(t) \propto (t/t_0)^{\frac{2}{3(1+w)}}$. Thus we must have

$$L(t) > c(t/t_0)^{\frac{5+9w}{3(1+w)}},$$

[21]Recall that for our purposes, observers are large quantum systems, well described by quantum field theory in a classical space-time, which have many semi-classical observables.

in order for the normal region to survive.[22] This is impossible. Indeed even to support a static interface between a $p = \rho$ fluid and a less stiff substance, we need a boundary stress tensor with equation of state parameter $w_s > 1$, violating the energy conditions, and causality.

For $w = -1$, we employ a different strategy. The asymptotic de Sitter cosmological horizon is a marginally trapped spherical null surface. Generic flat FRW models, and the $p = \rho$ geometry in particular, have black hole solutions of arbitrary horizon radius, so we can join a de Sitter space interior to such a black hole exterior. The Israel conditions are satisfied without violating energy conditions.

We will use these results in two ways. Recall that we are trying to find the most probable way (with a flat measure on the states of our system) to avoid the DBHF and create a long lived normal universe. Combining the above results with the conjecture [20] that de Sitter space is a system with a finite number of quantum states, we see that this will be achieved if the normal region asymptotes to de Sitter space. Note further, that once the asymptotic de Sitter situation is reached, the entropy of the system is the same as it would be if it were a homogeneous DBHF. This may be another way to understand the stability of the asymptotic dS sphere. Thus, the low entropy of the state of a DBHF with a normal, asymptotically dS defect, is a temporary phenomenon. Once the universe evolves to its dS vacuum state, the defect has the same entropy as an equivalent region of DBHF.

From this point of view, *the cosmological constant is an initial condition.* We can imagine a meta-verse consisting of the infinite DBHF seeded with a variety of asymptotically dS defects of varying horizon sizes. Given an *a priori* flat probability measure on the states of the system, larger values of the c.c. are preferred. The c.c. is thus a random initial condition, taking on the largest value compatible with whatever our prior conditions are for the "existence of a long lived normal universe". These might include the criteria of Weinberg [22] and/or Bousso [23], as well as others, which are determined by the manner in which the parameters of local physics depend on the c.c.[20, 24].

These considerations show that the early universe, prior to its asymptotic de Sitter phase, cannot begin as a homogeneous spherical defect filling up the coordinate volume of the eventual cosmological horizon. Rather we view it as a more complicated shape, pictorially modeled as a "tinker toy" made by joining together normal spheres, as shown in Fig. 4.

The optimal shape of this region cannot be determined without a more mathematical understanding of the quantum theory. It is the most probable shape conditioned on survival of the normal region. We will see some *a posteriori* constraints on it below. For the moment, we will need only one parameter, the fraction ϵ of the spatial volume of the initial value surface pictured in Fig. 4, which is taken up by the normal defects.

[22] Note that if $w = 1$, we do not need this stringent inequality and we satisfy all conditions with $L(t) = L$, a constant. The point is that here we simply have the $p = \rho$ geometry everywhere and the horizon is outside the artificial sphere we have drawn.

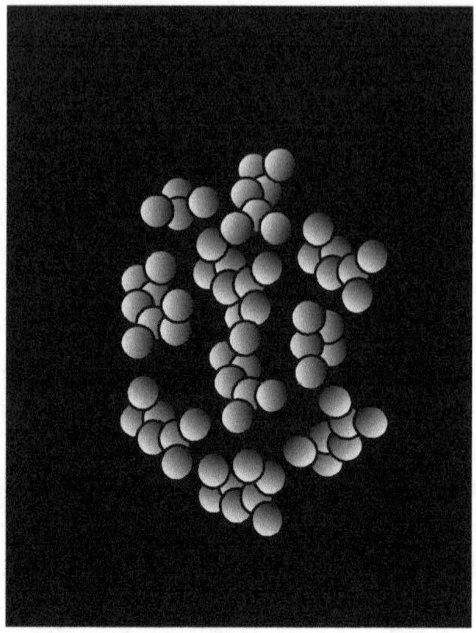

Fig. 4 Tinker–Toy of normal regions on the initial slice.

Let us pause for a moment to define what we mean by the initial value surface. This is the surface on which it first becomes reasonable to make the coarse grained geometrical approximation to the microscopic model of random Hamiltonians given in the previous section. Roughly speaking it is the value of k, the number of area steps away from the Big Bang, at which the Wigner semi-circle law becomes a good approximation to the spectral density. However, we must also insist that the effect of random irrelevant perturbations to the free fermion model is sufficiently suppressed.

The DBHF provides a background coordinate system, the FRW coordinates of the flat $p = \rho$ geometry in which we can define a density of normal defects, $\rho_N(\mathbf{x}, t)$. Note that, as in conventional cosmological discussions, this function is defined everywhere inside the coordinate patch that will eventually become the cosmological horizon of the asymptotically de Sitter normal cosmology. Its definition, in a hypothetical mathematical model of this holographic cosmology, would involve both the degrees of freedom inside the particle horizon, and those outside it. This does not violate causality, since the dynamics is designed to allow interactions only between degrees of freedom inside the horizon at any time. Rather, this global correlation reflects the necessity of choosing special initial conditions for *all* the degrees of freedom which will eventually come into the horizon, in order to avoid collapse into the DBHF. Our Israel junction condition arguments indicate the necessity for such global constraints on the initial conditions.

The tinker-toy of normal regions contains interstices which are filled with DBHF. The density of these regions is determined by the shape of the tinker-toy, according to

$$\rho_{BH}\mathbf{x},\mathbf{t} = \int ds \int d^3y \; f(\mathbf{x} - \mathbf{y}, t, s)\rho_N(\mathbf{y}, s).$$

The transfer function f is translation invariant, because the translation invariance of the DBHF is broken only by ρ_N. We will make one further assumption about the transfer function, namely that the correlations are short range, in the sense that the Fourier transform of f goes to a time independent constant at small $|\mathbf{p}|$. This seems evident from looking at pictures of possible tinker-toy shapes.

We can now use the scale invariance of the large k DBHF to constrain the two point function of $\rho_N(\mathbf{x}, t)$ during the $p = \rho$ era. Scale invariance manifests itself in the geometry in the form of the conformal Killing symmetry:

$$t \to bt; \quad \mathbf{x} \to b^{2/3}\mathbf{x}.$$

Every flat FRW geometry with a single component equation of state has such a conformal symmetry, but in most cases this is *not* a symmetry of the dynamics of the fluid. In the case of the DBHF, our explicit Hamiltonian becomes invariant under the symmetry for large k. Thus, although we don't really know the dynamics of the density ρ_N, the probability $\int d^3x \; \rho_N$ should be invariant under the scaling symmetry. Thus

$$G(k, s, s^{prime}) \equiv \langle \rho_N(\mathbf{k})\rho_N(-\mathbf{k})\rangle = G(b^{-\frac{2}{3}}k, bs, bs').$$

Similar formulae hold for higher point functions.

In the absence of a mathematical model for the tinker-toy fluctuations, we cannot prove that the fluctuations are Gaussian. One can invent any distribution for the initial fluctuations. However, we will see below that in order to insure that the normal part of the universe does not re-collapse into a DBHF, we must assume that all of the fluctuations are small. The simplest way for this to happen, would be if the effective probability distribution for ρ_N had the form

$$P = e^{-\frac{1}{\delta^2}F[\rho_N]}.$$

In this case the non-Gaussian fluctuations would be sub-leading. If these fluctuations have something to do with the CMB, then $\delta \sim 10^{-5}$, and the non-Gaussian fluctuations would be very small.

Now we come to the key observation which allows a normal universe to exist. Recall that we are *assuming* that expansion in the normal region of space-time is not affected by the interstitial DBHF. Let ϵ be the fraction of the coordinate volume of the initial[23] time slice, which is covered by normal regions. We expect ϵ to be a small number, because we are looking for the highest entropy configuration consistent with the emergence of a large normal universe. Taking ϵ close to one would mean that we are assuming low entropy homogeneous isotropic radiation as the initial state in most of our present cosmological horizon volume.

[23] Initial refers to the time at which the microscopic discrete time dynamics of the DBHF is well approximated by its coarse grained description as a scale invariant $p = \rho$ FRW universe.

Fig. 5 After the transition to dilute black hole gas.

On equal area time slices, the physical volume of a given coordinate volume of normal region grows like $t^{(3/2)}$,[24] while in the DBHF regions, physical volume grows like t. Thus, in a time of order $\frac{1}{\epsilon^2}$ a better picture of the coordinate volume of the cosmological horizon, is given by Fig. 5.

Most of the volume of the universe is covered by normal fluid. The interstitial DBHF now becomes localized regions of maximal entropy, in other words, isolated black holes. The region inside the eventual cosmological horizon looks like a radiation filled universe with a distribution of localized black holes. We say that at this time the universe has made the transition from the DBHF to a *dilute black hole gas*. If T is the size of the horizon at this phase transition then the energy density is $\sim \frac{1}{T^2}$.

Indeed, the black holes have a typical mass of order T, the size of the horizon at the time this transition takes place. The horizon scale t_0 at which the DBHF can be well described as a $p = \rho$ fluid is somewhat larger than the Planck scale and $T \sim \frac{1}{\epsilon^2} t_0 >> M_P^{-1}$. Thus, the black holes are relatively stable, with an evaporation time $\sim T^3$. If the black hole gas is approximately homogeneous, it will soon come to dominate the energy density of the universe. The initial radiation in the normal regions is negligible by comparison.

---

[24]We are here assuming the normal region is radiation dominated. This is motivated by an heuristic picture in which a normal region contains a black hole smaller than the horizon size, which decays into radiation. Perhaps a better argument is that, while our definition of normal regions is meant to imply that the system in these regions is described by effective field theory, we are at a high enough energy density to expect the relevant degrees of freedom to be described by some conformal field theory.

Now suppose that there are inhomogeneous fluctuations in the black hole positions and/or velocities. Obviously, this will cause the black holes to attract each other, collide, and form larger black holes. Since we are very close to the phase transition between DBHF and dilute gas, these collisions are very dangerous. They tend to return the system to the DBHF state. Even small fluctuations can be dangerous, because the homogeneous equation of state is now $p = 0$ and the fluctuations grow like $(t/T)^{2/3}$. One possible way to avoid this disaster is for the black holes to decay to radiation before the fluctuations grow to be of order one. The condition for this is

$$\frac{\delta\rho}{\rho}(T^3/T)^{2/3} \ll 1.$$

This seems to define a possible universe according to holographic cosmology, and it was the path followed in [12], before many of the details of this system were understood. However, it is not the universe we live in. The detailed question of what happens to the fluctuations generated during the $p = \rho$ era depends on the nature of the low energy effective field theory that describes the world, including the strength of interactions in the radiation gas, the nature of dark matter, etc. However, nothing can change the fact that the longest wavelength of these fluctuations is the horizon size T, at the time of transition, scaled up by the expansion of the universe. This is much smaller than the current horizon size, so these fluctuations could not be identified with what is seen in the CMB. Furthermore, even though we will see that the fluctuations are scale invariant, they do not spend a long period with wavelength larger than the Hubble horizon, and so there is no reason to expect them to be oscillating in phase. Thus, even if they had the right wavelength, they would not explain the shape of the CMB curves.

It is well known that these problems are solved by inflation, and that there are no other universally accepted solutions.[25] Holographic cosmology presents a new take on the question of whether inflationary initial conditions are generic. Low energy effective field theory (in the presence of a dilute gas of black holes) becomes a valid description of holographic cosmology, when the universe has expanded by a factor $A > 1$, beyond the point we have called the phase transition between DBHF and dilute gas. We don't have the tools to estimate A at this point but a reasonable guess is between 5 and 10. It is completely plausible that the appropriate low energy field theory has fields in it with a Lagrangian of the form

$$\mathcal{L} = \sqrt{-g}\Big[\frac{1}{2}G_{ij}(\phi)\nabla\phi^i\nabla\phi^j - \mu^4 V(\phi)\Big].$$

Note that we have taken the natural scale in field space to be the reduced Planck mass, 2×10^{18} GeV. General arguments [25] suggest that the potential and metric

[25] I am here classifying pre-Big Bang and cyclic universe scenarios among those not universally accepted.

on this scale will have no very small coefficients. It might vary on some much smaller scale (corresponding to large coefficients), which is generated by dynamical mechanisms. For such a Lagrangian the general expectation is that the number of e-folds $N_e \sim 1$[26] and the density fluctuations produced by quantum fluctuations of the inflaton field are $\sim \sqrt{N_e}\mu^2$, where we recall that all formulae are in reduced Planck units. Special forms of the Lagrangian [26, 27] can naturally accommodate large deviations from the latter formula.

Holographic cosmology provides a rationale for why the initial conditions for these scalar fields, on the time slice when field theory becomes a reasonable approximation, are homogeneous on the scale of the eventual cosmological horizon. Just like the fluctuations in the dilute black hole gas, large inhomogeneous fluctuations would lead to black hole collisions and growth, which precipitate re-collapse to the DBHF. *This is the holographic answer to the Boltzmann-Penrose question of why the universe appeared to start off in a very low entropy homogeneous state, giving rise to a thermodynamic arrow of time.*[27] *The generic initial state for holographic cosmology is the DBHF, in which no sort of complex organization of degrees of freedom exists. Our claim is that the most probable initial conditions which do permit complex organization, lock up their entropy in a dilute almost-homogeneous gas of black holes.* This leads to a rationale for the homogeneous initial conditions needed for inflation,[28] and suggests that inflation is one of only two mechanisms for avoiding collapse to the DBHF.

Note that in this scenario for inflation, homogeneity, isotropy and flatness are guaranteed by a non-inflationary mechanism: the combination of these properties of the DBHF, with the requirement of small fluctuations to avoid re-collapse. Inflation's role is to explain the length scale of fluctuations in the CMB, and their synchronization. Inflationary quantum fluctuations *may* also be the origin of the fluctuations we see in the CMB.

Indeed, in this model there are two possible sources of fluctuation in the universe. The condition for inflation to begin is that at some FRW time $t > AT$

$$\rho_{BH} \sim \frac{1}{t^2 T^2} = \mu^4.$$

---

[26]So that $N_e = 10-20$ is plausible, $N_e \sim 60-100$ seems to require some explanation, and much larger N_e seems terribly unlikely.

[27]Note that holographic cosmology has a built in arrow of time, coming from the direction in which the number of degrees of freedom inside the horizon increases. However, in the DBHF solution, there is no thermodynamic arrow of time since the system inside the horizon is always in equilibrium.

[28]Inflation theorists often argue that there is no need for homogeneity over our current horizon volume, but only over a small patch of order a few times the inflationary Hubble scale. The argument deals with those degrees of freedom not describable by field theory in that initial patch (but which are so describable today) by invoking the adiabatic theorem and assuming those degrees of freedom are in their ground state. But the adiabatic theorem is only valid for very special states of large quantum systems, so this is tantamount to *assuming* a particularly low entropy starting point for the universe.

Note that, in Planck units, $\mu^4 A^2 T^4 \leq 1$. It does not make sense to postulate a μ larger than this bound, because the effective field theory description of the universe is not valid at such large energy densities. The conventional approach to effective field theories of cosmology would allow us to choose any μ less than the Planck scale, however this is simply not allowed in holographic cosmology.

Inhomogeneities in ρ_{BH} now translate into inhomogeneities in the classical initial conditions of the inflaton, according to the equation

$$\mu^4 V'(\phi_0)\delta\phi = \delta\rho_{BH},$$

where

$$\mu^4 V(\phi_0) = \langle\rho_{BH}\rangle.$$

Thus, inflationary quantum fluctuations must be superimposed on the classical fluctuation in the inflaton initial condition imposed by the primordial inhomogeneity of the universe.

Above we showed that the two point function of the black hole density fluctuations is (for long wavelengths)

$$\langle\rho_{BH}(p,T)\rho_{BH}(-p,T)\rangle \sim \int_{t_0}^{T} ds\, ds' G(p,s,s'),$$

where t_0 is the time at which the FRW description of the DBHF becomes valid. Let $p_0 \equiv t_0^{-1}$. Choosing $b = (p/p_0)^{-\frac{3}{2}}$, and changing variables to bs and bs' in the integrals, we get

$$\langle\rho_{BH}(p,T)\rho_{BH}(-p,T)\rangle \sim (p/p_0)^{-3} \int_{(p/p_0)^{\frac{3}{2}} t_0}^{(p/p_0)^{\frac{3}{2}} T} ds\, ds' G(1,s,s'),$$

The lower limit on these integrals is effectively zero, for wavelengths greater than a few times t_0. If we assume that the correlation function G falls off exponentially at large times (as would be the case for the propagator of any sort of massive or massless field), then the integral is p independent as long as $(p/p_0)^{3/2}T \gg t_0$. Since the integral is falling exponentially, the $\gg$ symbol in this equation just means a factor of two or so. Thus the density fluctuation spectrum generated during the DBHF fluid era is scale invariant over a fixed range of scales and vanishes outside that range. Recalling that p is a coordinate wave number, and that physical length scales have expanded by $(T/t_0)^{1/3}$ the longest wavelengths for which we have scale invariance, have physical size of order $R_0 \sim T/2^{2/3}$, which is of order the horizon scale at the time of transition to the dilute black hole gas.

2.3 Phenomenology

We are considering a model where inflation starts at an average energy density

$$\mu^4 = \frac{1}{A^3 T^2},$$

where $A \geq 10$ is the expansion of the scale factor between the phase transition to a dilute black hole gas, and the point at which we can make a reliable effective field theory analysis. The magnitude of primordial fluctuations is constrained by[29]

$$\frac{\delta\rho_{BH}}{\rho_{BH}} A < 1,$$

in order to avoid re-collapse to the DBHF. In what follows, it will be important to distinguish restrictions on parameters, which follow from consistency of the model, from phenomenological requirements, and this bound is a consistency condition.

We also note that our philosophy is that t_0, ϵ and μ[30] are parameters that we could in principle determine completely if we had a real mathematical model of the cosmology we are discussing. However, at present we can only derive rough bounds on these parameters following from consistency of the picture, and then further constrain them by phenomenology. The phenomenological constraints will depend on our assumptions about how the model is related to observation. Apart from the restriction on fluctuations, the *a priori* consistency conditions are $t_0 \sim 10$, $\epsilon < 0.1$, and $A \sim 10$. Note that A is defined in terms of T and μ so this is a constraint on those two parameters.

We have shown that there are scale invariant fluctuations on a range of physical scales (at the time of the phase transition) between t_0 and T. We will first consider the possibility that these fluctuations are related to what is seen in the CMB, and then examine the alternative hypothesis of inflationary quantum fluctuations. First we note that the beginning of inflation is determined by the local black hole density, so inhomogeneities of ρ_{BH} on the equal area time slices, translate into inhomogeneities in the inflaton field via the equation

$$\delta\rho_{BH}(\mathbf{x}, t_I) = \mu^4 V'(\phi_0(t_I))\delta\phi(\mathbf{x}, t_I) + \dot{\phi}_0 \delta\dot{\phi}(\mathbf{x}, t_I),$$

where $\phi_0(t)$ is the homogeneous inflaton solution.

The first requirement of a successful model is that the range of scales over which the DBHF gives scale invariant fluctuations, be large enough to include what is observed. This means 10^3 if we only want to explain the CMB and 10^5 if we want

---

[29]Careful readers of previous work will note discrepancies between the present section and those earlier estimates. I believe the current discussion is the more accurate one.

[30]And the number of e-foldings of inflation, N_e.

approximately scale invariant fluctuations down to galactic scales. We will add an extra order of magnitude to the range of scales required, in order not to have to rely on a precise fine tuning of the amount of inflation that occurs. Thus

$$T/t_0 \geq 10^4 - 10^6.$$

This corresponds to

$$\epsilon \sim 10^{-2} - 10^{-3},$$

for the underlying volume fraction parameter. This seems quite reasonable. Recall also that we have estimated $t_0 = 10m_p^{-1}$, where $m_p = 2 \times 10^{18}$ GeV is the reduced Planck mass. We will work in units where $m_P = 1$.

The current value of the largest scale at which we can expect to see approximately scale invariant fluctuations is

$$R = TAe^{N_e} L(T_{RH}/T_{NOW}) = TAe^{N_e} L(\mu^3/T_{NOW})$$

.[31] L is the expansion of the scale factor during the matter dominated period between the end of inflation and reheating, and we have made the standard estimate of T_{RH}, the reheat temperature, in terms of the inflaton Lagrangian parameter μ. Similar estimates give $L = \mu^{-4}$. Plugging in $T_{NOW} = 1.2 \times 10^{-31}$, and requiring that R be at least the current horizon size. 5×10^{61}, we obtain

$$TAe^{N_e}\mu^{-1} = e^{N_e} T^{3/2} A^{7/4} \sim 10^{30}.$$

μ is bounded from below by the requirement that reheating occur at a temperature higher than nucleosynthesis $\mu^3 > 6\text{MeV} = 3 \times 10^{-21}$. In terms of A and T, this is $3T^{3/2}A^{7/4} < A^{-1/2} \times 10^{21}$. So we get a bound on the number of e-foldings

$$e^{N_e} \geq 3 \times 10^9 A^{1/2} > 10^{10}.$$

Holographic cosmology needs only about 20 e-folds of inflation to explain the universe we see, under the assumption that the CMB fluctuations arose during the DBHF era.

Note that the phenomenological bound on T and the requirement $A \geq 10$ put an upper bound on μ

$$\mu^4 \leq 10^{-13} \to 10^{-17},$$

depending on whether we want to explain the fluctuations required for galaxy formation. For the maximal value of μ, obtained when $A = 10$, and the primordial black hole fluctuations are seen in the CMB but not at galaxy scales, the required number of e-folds is 42. The reheat temperature is of order 10^{11} GeV.

[31] We have neglected entropy dumps which occur later than the primordial reheating of the universe.

In this model, variations in μ within the phenomenologically allowed range are correlated with variations in A, and thus with the *a priori* bound on the strength of primordial fluctuations. Note however that we will not be able to obtain an argument which explains why the observed CMB fluctuations are as small as they are, for any value of μ. We have seen that primordial fluctuations in the black hole density are converted into fluctuations in the initial conditions for the inflaton field. Those fluctuations which have physical wavelengths longer than the inflationary Hubble scale will not change their amplitude as the universe inflates. Our bound on the primordial fluctuations comes from insisting that their amplitude is small enough for inflation to actually occur. I see no reason why they should be smaller than say 10^{-2} *at the time when inflation begins*. If μ is small, this implies a much more dramatic bound on the amplitude of fluctuations at the dense to dilute black hole phase transition.

Now let us consider the possibility that the fluctuations in ρ_{BH} are, for some reason, negligible. I believe that a plausible constraint on the volume fraction on the initial surface, ϵ, of the asymptotic dS cosmological horizon coordinate volume, which is occupied by normal regions, is $\epsilon < 0.1$.[32] This translates into a bound $T > 10^3$, whence $\mu^4 < 10^{-9}$. Thus, we can build a conventional slow roll inflation model, which adequately explains fluctuations on both the CMB and galactic scales, as long as $e^{N_e} > 10^6$. This is in some sense the most conservative kind of holographic cosmology, but it is somewhat disappointing because there is no direct observational evidence of the primordial state of the universe. However, even in this scenario, the primordial state *does* serve a purpose. It explains why inflationary conditions are plausible, solving the Boltzmann-Penrose conundrum, and as a consequence, it relieves some of the model-building tension on inflationary models. Inflation is now required only to explain the nature of CMB and galaxy seed fluctuations, and we can make do with fewer e-folds. The only puzzle I see in these models is why the black hole fluctuations are so small. Our working hypothesis has been that they are as large as they can be without causing re-collapse into the DBHF, in order for our universe to qualify as a generic product of collapse avoidance. The simplest resolution of this puzzle is that the black hole fluctuations are simply expanded to large scale by inflation. Indeed, the parameters $\epsilon \sim 0.1$ and $t_0 \sim 10$ which allow for conventional high scale inflation, give black hole density fluctuations over only two orders of magnitude in scale. Today these are the scales $e^{N_e}(10^{38} \to 10^{40})$. So if $e^{N_e} > 10^{23}$, these scales are outside our horizon.

2.3.1 Spectral Tilt and Over the Top Inflation

In earlier work on holographic cosmology [28], Fischler and I assumed that the scale invariant black hole fluctuations simply translated into scale invariant fluctuations of

[32] Recall that we are trying to find the most probable initial conditions, which can lead to a universe that escapes collapse into the DBHF.

the inflaton energy density via the equation that determines the beginning of inflaton dominance of the energy density,

$$\rho_{BH} + \delta\rho_{BH} = \frac{1}{2}\dot{\phi}^2 + \mu^4 V(\phi).$$

As a consequence we declared that holographic cosmology predicted an *exactly* scale invariant spectrum. Current data is marginally inconsistent with that prediction, and so we retreated to the conservative models described in the previous paragraph.

Some years ago, in unpublished work, Fischler and I realized that we had left something out, and that our model *could* produce a spectral tilt. However, for conventional inflation scenarios, it seemed to produce a tilt in the wrong direction, towards the blue. Since we could eliminate the tilt by playing with the potential, there didn't seem to be much point in publishing. In the course of preparing these notes, I realized that there *might* be a way to obtain a red tilt in some part of the spectrum. The details of this are not worked out, and it is not clear that these models make sense, but I include a brief description of the idea here for completeness.

Let us expand the initial condition equation around the homogeneous background, recalling that in holographic cosmology we have a principled reason for insisting that the fluctuations are small. We get a fluctuating inflaton energy density

$$\frac{\delta\rho_I}{\rho_I} = \frac{\dot{\phi}_0\delta\dot{\phi} + \mu^4 V'(\phi_0)\delta\phi}{E[\phi_0]},$$

where ϕ_0 is the homogeneous background and E its energy. We see that the black hole fluctuations induce fluctuations in the initial conditions for the inflaton field.

The interesting questions for observation is the magnitude of each mode of fluctuation as its scale becomes larger than the inflationary horizon. These are the quantities that are conserved during the evolution outside the horizon. Since they exit the horizon during inflation, the velocity of the homogeneous field is negligible and we have, for the energy density fluctuations at horizon exit

$$\frac{\delta\rho(k)}{\rho} = \frac{V'(\phi_0(t_k))}{V(\phi_0(t_k))}\delta\phi(k).$$

t_k is the time at which the mode k exits the horizon. $\delta\phi(k)$ is a random variable with scale invariant two point function. However, there is additional k dependence coming from the motion of the inflaton and the variation of the logarithmic derivative of the potential. This is similar to what happens for inflationary quantum fluctuations, but the formulae differ in detail, because the coefficient of the two point function does not change with k.

Here now is the apparent disaster. For a conventional inflationary scenario, the logarithmic derivative is increasing with time, because the inflaton is rolling down a potential. Modes that go out of the horizon later, have larger amplitudes. But the usual "last out, first in" rule for inflation tells us that the later modes are those of shorter wavelength. So this would lead to a blue tilt of the spectrum. The best we can do is to choose a potential with slowly varying logarithmic derivative, which leaves us with a scale invariant spectrum.

However, it is possible that a different scenario for the inflaton motion can give us a more interesting spectrum. Suppose that at the initial momentum of inflaton dominance, the inflaton is traveling up toward a maximum of the potential, not yet in friction dominated motion, but that the motion becomes friction dominated, and inflation begins, as the inflaton struggles up a reasonably steep hill. Suppose further that the potential to the right of the maximum is much less steep. During the initial, climbing phase of inflation, the logarithmic derivative of the potential is increasing. Thus, among modes that exit the horizon during this phase there will be a red tilted spectrum. The tilt will gradually go away and then be replace by a less dramatic blue tilt. The whole modulation must be multiplied against a scale invariant spectrum with cutoffs, and the true spectrum will depend on where the cutoffs sit relative to the modulation.

I think that, given the quality of the data, there is little doubt that one can cook up a model like this, which fits the data. The real question is whether the model is robust and theoretically plausible, or involves fine tuning. I don't know the answer to those questions at present, so I will not claim that there is a version of holographic cosmology that gives a red-tilted spectrum, without invoking inflationary quantum fluctuations.

2.3.2 Executive Summary

To summarize the phenomenological prospects of holographic cosmology:

- Holographic cosmology explains the homogeneity, isotropy, flatness, and low entropy initial conditions for the universe in terms of a mathematical model called the DBHF, and hypothetical variations of it which describe a "normal tinker-toy" embedded in the dense black hole fluid. A number of properties of this tinker-toy universe can be established using the Israel Junction condition and simple considerations about black holes and the growth of fluctuations. In particular, it *must* evolve to an asymptotically de Sitter universe, with a cosmological constant that is determined by initial conditions.
- The highest energy density scale at which this universe can be described by effective field theory is at most of order 10^{-6} in reduced Planck units. It might be much lower if it turns out that the parameter t_0 must be larger than 10, or, more plausibly, the volume fraction, ϵ of normal fluid on the initial data surface must be < 0.1. This is actually required by phenomenology if we want the density fluctuations generated by the primordial dynamics to leave an imprint in the CMB. Just below this scale, the universe is described as a gas of black holes, with

almost uniform density. The density is high and small fluctuations will make the system revert to the DBHF.

- Indeed, if no inflation occurs, then one can derive an *a priori* bound

$$\frac{\delta\rho}{\rho} 10^4 \ll 1,$$

where the factor of 10^4 might well be larger if t_0 is larger or ϵ smaller. This version of the model might well resemble our universe in a coarse grained way, but could not have correlations in the CMB that span the whole sky, nor an explanation of the synchronized acoustic oscillations which are evident in the CMB data.
- If inflation occurs, the bound on primordial black hole density fluctuations depends on the scale of inflation. Consistency of the model requires that this scale be less than $10^{-2} \rightarrow 10^{-3}$, but there are versions of the model with μ as low as 10^{-7}. The low μ versions have $\epsilon \sim 10^{-3}$ and $t_0 \sim 10$. The reheat temperature is close to nucleosynthesis temperature, which has profound implications for particle physics. SUSY dark matter does not work. The most plausible dark matter candidate is a QCD axion with decay constant $\sim 10^{14} \rightarrow 10^{15}$ GeV. Baryogenesis proceeds most plausibly via a low scale Affleck–Dine mechanism, or through the decay of the inflaton, but the latter mechanism requires the existence of renormalizable baryon violating couplings in the MSSM.
- There is a more conservative version of the model in which inflation takes place at about 10^{15} GeV. CMB and galaxy seed fluctuations are explained as inflationary quantum fluctuations, in the usual manner. Fluctuations in the black hole density are much larger than these in magnitude, but take place over only two orders of magnitude in scale size. The number of e-foldings is such that this range of scales is invisible at present. This version has $t_0 = 10$ and $\epsilon = 0.1$.
- The radical and conservative versions of the model can in principle be distinguished by the tilt of the CMB spectrum. A relatively new analysis shows that the radical version gives a blue tilted or scale invariant spectrum for any conventional inflaton trajectory. It might give a more complicated spectrum, red tilted at the IR end, and blue tilted on the UV side (the scale where galaxies are formed), in a more complicated inflation scenario, dubbed *over the top inflation*. The conservative version of the model gives the usual inflationary prediction of a red tilted spectrum.

2.4 The Thermodynamic Arrow of Time

Holographic cosmology addresses, but has not yet resolved the question of the thermodynamic arrow of time. Note that while the holographic formulation of cosmology is *not* invariant under TCP, this does not by itself explain the thermodynamic arrow of time. The DBHF is not a TCP invariant system. Its time direction is correlated with the increase in the number of degrees of freedom which are allowed to interact with each other by its time dependent Hamiltonian. However, at each instant, the state of the interacting degrees of freedom is a random state in the available Hilbert space. Thus, although there is a global sense in which the entropy of the system increases with time, there is no sense in which the entropy visible to a localized observer[33] increases. It is always at the maximum consistent with the covariant entropy bound.

Our heuristic tinker-toy cosmology *does* have a thermodynamic arrow of time. The initial localized degrees of freedom are a gas of black holes, very close to the density at which they would coalesce and form the DBHF phase. If the gas is non-uniform (on the homogeneous equal area slices whose existence is a consequence of the primordial DBHF) it will re-collapse into the dense phase. The survival of normal regions of space-time, describable by effective field theory, is impossible unless the state of the localizable degrees of freedom is quite homogeneous. Thus, the homogeneity, isotropy, and low entropy initial conditions for the localizable degrees of freedom in the observable universe are conditions for survival of any normal region of the universe.[34] This argument is satisfactory on a qualitative level, but we do not yet have a quantitative argument for the size of primordial fluctuations away from homogeneity.

A model which does provide such an estimate is the original version of holographic cosmology, with no inflation at all. In this model, the Big Bang originates from the decay of a nearly homogeneous gas of black holes. A bound on the size of primordial fluctuations follows from the requirement that the decay occurs before fluctuations in the density of black holes go non-linear. The bound would be saturated for generic initial conditions. This model has only one parameter, the horizon size T at the time of transition between DBHF and dilute black hole gas. This parameter would be calculable in a full mathematical treatment of the DBHF-defect cosmology. However, this model is not compatible with either the wavelengths of the CMB fluctuations, or the synchronization of their initial velocities, which leads to the observed acoustic peaks.

A fully quantitative explanation of the thermodynamic arrow of time, which predicts the size of primordial fluctuations, will have to wait for a more detailed mathematical formulation of the DBHF-defect cosmology.

Acknowledgements This research was supported in part by DOE grant number DE-FG03-92ER40689.

---

[33] In this bizarre context, a localized observer simply means a subsystem of the states available in a causal diamond at any time.

[34] The large scale homogeneity and isotropy and flatness of the DBHF also play a role in this argument.

References

1. H. Lindblad, I. Rodnianski, *The Global Stability of the Minkowski Space-Time in Harmonic Gauge*, (2004), p. 59 [e-Print: math/0411109]; H. Ringstrom, On curvature decay in expanding cosmological models, Commun. Math. Phys. **264**, 613–630, (2006)
2. P.P. Kulish, L.D. Faddeev, Theor. Math. Phys. **4**, 745 (1970) [Teor. Mat. Fiz. **4**, 153 (1970)]
3. O. Aharony, S.S. Gubser, J.M. Maldacena, H. Ooguri, Y. Oz, Phys. Rept. **323**, 183 (2000) [arXiv:hep-th/9905111]
4. T. Hertog, G.T. Horowitz, J. High Energ. Phys. **0407**, 073 (2004) [arXiv:hep-th/0406134]
5. R. Penrose, unpublished; P.D. D'Eath, P.N. Payne, Phys. Rev. D **46**, 658 (1992); H.J. Matschull, Class. Quant. Grav. **16**, 1069 (1999) [arXiv:gr-qc/9809087]; T. Banks, W. Fischler, arXiv:hep-th/9906038; D.M. Eardley, S.B. Giddings, Phys. Rev. D **66**, 044011 (2002) [arXiv:gr-qc/0201034]
6. T. Banks, Lecture at the 60th Birthday celebration for David Gross, KITP, UCSB, 2001
7. O. Aharony, T. Banks, J. High Energ. Phys. **9903**, 016 (1999) [arXiv:hep-th/9812237]
8. N. Seiberg, Nucl. Phys. Proc. Suppl. **67**, 158 (1998) [arXiv:hep-th/9705117]
9. O. Aharony, M. Berkooz, S. Kachru, N. Seiberg, E. Silverstein, Adv. Theor. Math. Phys. **1**, 148 (1998) [arXiv:hep-th/9707079]
10. R. Bousso, A covariant entropy conjecture. J. High Energ. Phys. **9907**, 004 (1999) [hep-th/9905177]; R. Bousso, Holography in general space-times. J. High Energ. Phys. **9906**, 028 (1999) [hep-th/9906022]
11. T. Banks, W. Fischler, *An holographic cosmology* [arXiv:hep-th/0111142]
12. J. Polchinski, [arXiv:hep-th/9901076]
13. L. Susskind, [arXiv:hep-th/9901079]
14. J. Madore, Class. Quant. Grav. **9**, 69 (1992)
15. F.J. Murray, J. von Neumann, Ann. Math. **37**, 116 (1936); Trans. Amer. Math Soc. **41**, 208 (1937); Ann. Math. **44**(2), (1943, MR5, 101; A. Connes, *Noncommutative geometry, Chapter V* (Academic Press, London, 1990)
16. S.R. Coleman, J. Mandula, Phys. Rev. **159**, 1251 (1967)
17. T. Banks, *II_∞ factors and M-theory in asymptotically flat space-time,* [arXiv:hep-th/0607007]
18. V.G. Lapchinsky, V.A. Rubakov, Canonical quantization of gravity and quantum field theory in curved space-time, Acta Phys. Polon. B **10**, 1041 (1979); T. Banks, W. Fischler, L. Susskind, Quantum cosmology in (2+1)-dimensions and (3+1)-dimensions. Nucl. Phys. B **262**, 159 (1985)
19. T. Banks, T C P, Quantum gravity, the cosmological constant and all that. Nucl. Phys. B **249**, 332 (1985)
20. T. Banks, *Cosmological breaking of supersymmetry or little Lambda goes back to the future. II* [hep-th/0007146]; *Cosmological breaking of supersymmetry?* Int. J. Mod. Phys. A **16**, 910 (2001)
21. W. Fischler, L. Susskind, *Holography and cosmology* [hep-th/9806039];
22. S. Weinberg, *Anthropic bound on the cosmological constant,* Phys. Rev. Lett. **59**, 2607 (1987)
23. R. Bousso, R. Harnik, G.D. Kribs, G. Perez, Phys. Rev. D **76**, 043513 (2007). [arXiv:hep-th/0702115]
24. T. Banks, *Cosmological supersymmetry breaking and the power of the pentagon: A model of low energy particle physics* [arXiv:hep-ph/0510159] T. Banks, *Remodeling the pentagon after the events of 2/23/06* [arXiv:hep-ph/0606313]
25. T. Banks, M. Dine, P.J. Fox, E. Gorbatov, J. Cosmology Astropart. Phys. **0306**, 001 (2003) [arXiv:hep-th/0303252]
    N. Arkani-Hamed, L. Motl, A. Nicolis, C. Vafa, J. High Energ. Phys. **0706**, 060 (2007) [arXiv:hep-th/0601001]

26. M. Alishahiha, E. Silverstein, D. Tong, Phys. Rev. D **70**, 123505 (2004). [arXiv:hep-th/0404084]
27. L. Randall, M. Soljacic, A.H. Guth, Nucl. Phys. B **472**, 377 (1996) [arXiv:hep-ph/9512439]
28. T. Banks, W. Fischler, L. Mannelli, *Microscopic quantum mechanics of the p = rho universe,* Phys. Rev. D **71**, 123514 (2005) [arXiv:hep-th/0408076]

The Emergent Nature of Time and the Complex Numbers in Quantum Cosmology

Gary W. Gibbons

Abstract The nature of time in quantum mechanics is closely related to the use of a complex, rather than say real, Hilbert space. This becomes particularly clear when considering quantum field theory in time dependent backgrounds, such as in cosmology, when the notion of positive frequency ceases to be well defined. In spacetimes lacking time orientation, i.e without the possibility of defining an arrow of time, one is forced to abandon complex quantum mechanics. One also has to face this problem in quantum cosmology. I use this to argue that this suggests that, at a fundamental level, quantum mechanics may be really real with not one, but a multitude of complex structures. I relate these ideas to other suggestions that in quantum gravity time evolution may not be unitary, possibly implemented by a super-scattering matrix, and the status of CPT.

1 Introduction

The topic of this volume is *The Arrow of Time*, but before considering that we should ask *What is the nature of time?*

Both Quantum Mechanics and General Relativity have something to say about this.

But what they say is not quite compatible

For example, in quantum mechanics, there may be observables or operators corresponding to spatial positions but time is not an observable, i.e. it is not an operator [117–119].[1] More precisely, by an argument going back to Pauli, commutation relations like

[1]See Pullin's contribution in this volume.

G.W. Gibbons (✉)
D.A.M.T.P., Cambridge University, Wilberforce Road, Cambridge CB3 0WA, UK
e-mail: gwg1@damtp.cam.ac.uk

L. Mersini-Houghton and R. Vaas (eds.), *The Arrows of Time*, Fundamental Theories of Physics 172, DOI 10.1007/978-3-642-23259-6_6,

$$\left[\hat{x}^{\mu}, \hat{P}_{\nu}\right] = i\delta^{\mu}_{\nu} \tag{1}$$

are incompatible with the spectrum of $\hat{p}^{\mu}$ lying in the future lightcone.

In General Relativity on the other hand, space and time are usually held to be on the same footing.

Because the nature of time in Quantum Mechanics is less familiar and less frequently discussed, than it is in General Relativity I shall begin by recalling [1, 8, 9, 11, 12, 16, 60] how *time is intimately connected with the complex* (Hilbert Space) *structure of quantum mechanics.*

In other words, the use of complex numbers and hence of complex amplitudes in Quantum Mechanics is intimately bound up with how Quantum States evolve in time.

$$i\frac{d\Psi}{dt} = H\Psi. \tag{2}$$

In particular there can be no evolution if Ψ is real.[2]

To proceed it is helpful to contemplate more deeply than is usual in cosmology.

2 The Structure of Quantum Mechanics

If one analyzes the *Logical Structure of Quantum Mechanics* one discovers that it consists of two different types of statements:

- **I** Timeless[3] statements about *states, propositions, the Principle of Superposition, probabilities, observables* etc.
- **II** Statements about how states and observables change, *Schrödinger's equation and Unitarity* etc.

The upshot of an analysis of Part **I** (so called Quantum Logic) [4, 5] is that pure states are points in a *Projective Space* over $\mathbb{R}, \mathbb{C}$ or $\mathbb{H}$[4].

$$\Psi \equiv \lambda\Psi, \qquad \lambda \in \mathbb{R}, \mathbb{C} \text{ or } \mathbb{H}. \tag{3}$$

Now any vector space over $\mathbb{R}, \mathbb{C}$ or $\mathbb{H}$ is a vector space V over $\mathbb{R}$ *with some additional structure* (cf. [1, 7]), so let's use real notation. Observables are *symmetric*

[2]Conversely, as shown by Dyson [3] in his three-fold way, if H is time-reversal invariant one may pass to a real (boson) or quaternionic (fermion) basis.

[3]Of course in splitting the discussion into two parts, in Part **I** we take the view that Quantum Logic like its classical Aristotelian special case is timeless. This avoids appealing to Temporal Logic to resolving such paradoxes as that of "the sea fight tomorrow" [72–74] and puts the burden of its resolution firmly where it belongs, in Part **II**.

[4]By the principle of binary coding, Classical Boolean Logic may, for finite sets at least, be thought of as projective geometry over the Galois field of two elements. We shall also ignore the exceptional case of the octonions.

bilinear forms:

$$\langle \Psi O \Psi \rangle = \Psi^a O_{ab} \Psi^b, \qquad O_{ab} = O_{ba}. \tag{4}$$

$a = 1, 2, \ldots, n = \dim_{\mathbb{R}} V$. Mixed states ρ are positive definite observables dual to the observables

$$\langle O\rho \rangle = \rho^{ab} O_{ab} = \mathrm{Tr}\,(\rho O), \qquad \rho^{ab} = \rho^{ba}. \tag{5}$$

There is a privileged density matrix *the completely ignorant density matrix* which we may think of as a *metric*[5] g_{ab} on V and use it to normalize our states

$$\langle \Psi | \Psi \rangle = g_{ab} \Psi^a \Psi^b, \qquad g_{ab} = g_{ba}, \qquad \mathrm{Tr}\rho = g^{ab} \rho_{ab}. \tag{6}$$

The upshot of a conventional analysis of **II** (Dirac called it *Transformation Theory*) is that states change by by means of linear maps which preserves the metric (i.e. preserves complete ignorance)

$$\Psi^a \to S^a{}_b \Psi^b, \qquad g_{ab} S^a{}_c S^b{}_d = g_{cd}. \tag{7}$$

Thus $S \in SO(n, \mathbb{R})$, $n = \dim_{\mathbb{R}} V$. Infinitesimally

$$S^a{}_b = \delta^a_b + T^a{}_b + \ldots, \tag{8}$$

where the endomorphism or *Operator* $T^a{}_b$ gives a *two-form* when the index is lowered

$$g_{ab} T^b{}_c := T^{\flat}_{ac} = -T^{\flat}_{ca}. \tag{9}$$

But Dirac taught us that, just as in Hamiltonian mechanics, *to every (Hermitian) Operator there is an Observable and vice versa.* How can this be? Our vector space V over $\mathbb{R}$ needs some extra structure, in fact a *complex structure* $J^a{}_b$ or *privileged operator* which also preserves the metric (i.e. preserves complete ignorance).

$$g_{ab} J^a{}_c J^b{}_d = g_{cd}. \tag{10}$$

Then

$$J^a{}_b J^b{}_c = -\delta^a_c \implies \omega_{ab} = -\omega_{ba}, \tag{11}$$

where the *symplectic two-form* $\omega_{ab} = g_{ac} J^c{}_b$ may be used to lower indices and obtain a symmetric tensor for every (Hermitian) observable (i.e. one that generates a transformation preserves the symplectic form)

$$\omega_{ab} T^b{}_c := T_{\flat ac} = +T_{\flat ca}. \tag{12}$$

[5]Strictly speaking the inverse.

We can think of this more group theoretically.[6] *In regular Quantum Mechanics* V is a Hermitian vector space its transformations should be unitary, but

$$U(\frac{n}{2},\mathbb{C}) = SO(n,\mathbb{R}) \cap GL(\frac{n}{2},\mathbb{C}), \tag{13}$$

where $GL(\frac{n}{2},\mathbb{C}) \subset GL(n,\mathbb{R})$ is the subgroup preserving J, and $SO(n,\mathbb{R}) \subset GL(n,\mathbb{R})$ is the subgroup preserving the metric g. One also has

$$U(\frac{n}{2},\mathbb{C}) = SO(n,\mathbb{R}) \cap Sp(n,\mathbb{R}), \tag{14}$$

where $Sp(n,\mathbb{R}) \subset GL(n,\mathbb{R})$ is the subgroup preserving the symplectic form ω, and of course

$$U(\frac{n}{2},\mathbb{C}) = Sp(n,\mathbb{R}) \cap GL(\frac{n}{2},\mathbb{C}). \tag{15}$$

2.1 A Precautionary Principle

Now the main message of this review is that given a vector space V over $\mathbb{R}$ *it may have no complex structure* (n must obviously be even!) or if it is does, *the complex structure may not be unique* (they are typically members of infinite families).

Thus on four dimensional Euclidean space $\mathbb{E}^4$ they belong (modulo a choice of orientation) to a two-sphere $S^2 = SO(4)/U(2)$.

More generally, every quaternion vector space has such a two-sphere's worth of complex structures,[7] i.e. *a two-sphere's worth of of times*!

To bring out the fact that in physics we use many different complex structures for many different reasons it is occasionally helpful to indicate explicitly by the symbol i_{qm} the very particular complex structure on the Hilbert space $\mathcal{H}_{\text{qm}}$ of the standard model and so that Schrödinger's equation really reads

$$i_{\text{qm}}\frac{d\Psi}{dt} = H\Psi. \tag{16}$$

At a more mundane level, the use of the notation i_{qm} brings out how dangerous and misleading, certainly to the beginner, it can be to use complex notation too sloppily. Suppose one has a theory, with an $SO(2)$ symmetry (gauged or un-gauged). It is tempting to collect the fields, e.g. scalars ϕ_1, ϕ_2 in pairs

$$\phi = \phi_1 + i\phi_2. \tag{17}$$

---

[6]Or recall what we might know about Kähler manifolds; Quantum Mechanics makes use of a Kählerian vector space.

[7]cf Hyper-Kähler manifolds such as K3.

Now the i, which generates the $SO(2)$ action is (17) is not the same as $i_{\rm qm}$. This is clear from the fact that charge conjugation

$$C : \phi_1 + i\phi_2 \to \phi_1 - i\phi_2 \tag{18}$$

is anti-linear, i.e anti commutes with i but is nevertheless represented on $\mathcal{H}_{\rm qm}$ as a linear operator, i.e. one which commutes with $i_{\rm qm}$. Note that there would be no temptation to indulge in such notational confusion if there were three scalar fields ϕ_1, ϕ_2, ϕ_3 and the symmetry $SO(3)$.

Just how confusing the sloppy use of the somewhat ambiguous complex notations currently can be is nicely illustrated in [34] in the context of quantum field theory, an example which will be of relevance later.

At a purely practical level, the avoidance of an excessive use of complex notation also helps in formulating action principles in an intelligible fashion. One is often instructed that in varying an action with, for example complex scalars, that one should vary the action regarding 'ϕ and its complex conjugate $\bar{\phi}$ as independent'. On the face of it this sounds ridiculous. What is actually meant is that one varies regarding the real ϕ_1 and imaginary ϕ_2 parts of ϕ as independent. It is easy to check that, as long as the action is real, then this cook book recipe will give the correct result, essentially because varying with respect to $\bar{\phi}$ gives the complex conjugate of the equation obtained by varying with respect to ϕ. However as a general principle the cook book recipe cannot be of general validity. It fails, and is inconsistent, if, for example, one varies a complex valued function of a complex variable and its complex conjugate. If it only works in special cases and can lead to incorrect results, it seems best to avoid both the cook book recipe and the misleading notation that gives rise to it.

In conclusion therefore, it seems wise to adopt a course of action, particularly at the classical level before quantization, in which one proceeds as far as possible by considering all physical quantities and their related mathematical structures to be real until one is forced to introduce complex notation and $i_{\rm qm}$ at the point where one introduces quantum mechanics.

In other words, in what follows, I plan to follow, in so far as is possible, Hamilton's course of action [27]

> The author acknowledges with pleasure that he agrees with M. CAUCHY, in considering every (so-called) Imaginary Equation as a symbolic representation of two separate Real Equations: but he differs from that excellent mathematician in his method generally, and especially in not introducing the sign $\sqrt{-1}$ until he has provided for it, by his Theory of Couples, a possible and real meaning, as a symbol of the couple (0, 1).

2.2 Dyson's Threefold Way

In this language, Dyson's observation [3] is that in standard quantum mechanics an anti-linear involution Θ acting on rays may be normalized to satisfy

$$\Theta^2 = \pm 1, \tag{19}$$

where the plus sign corresponds to an even spin state and the odd sign to an odd spin state.To say that Θ is anti-linear is to say that it anti-commutes with the standard complex structure $i_{\rm qm}$, ${i_{\rm qm}}^2 = -1$

$$\Theta i_{\rm qm} + i_{\rm qm}\Theta = 0. \tag{20}$$

Now for the plus sign Θ, is a projection operator and we get what is called a *real structure* on the original complex Hilbert space and if the Hamiltonian is time-reversal invariant, then we may use the projection operator to project onto the subspace of real states. On the other hand for the minus sign we construct

$$K = \Theta\, i_{\rm qm} \tag{21}$$

and find that Θ, $i_{\rm qm}$, K satisfy the algebra of the quaternions.

2.3 Relation to Jordan Algebras

There is an interesting tie in here with the theory of Jordan algebras [54, 55] which were originally introduced by Jordan as a possible avenue for generalizing quantum mechanics but in the end led to the same three basic possibilities.

In all three varieties of quantum mechanics the states, i.e. the space of positive semi-definite Hermitian matrices in $\mathrm{Herm}_n(\mathbb{K})$, $\mathbb{K} = \mathbb{R}, \mathbb{C}, \mathbb{H}$ form a homogeneous convex self-dual cone, Moreover as observed by Jordan, they satisfy an abelian, but non-associative, algebra whose multiplication law is

$$\big(O_1, O_2\big) \rightarrow \frac{1}{2}\big(O_1 O_2 + O_2 O_1\big). \tag{22}$$

The algebra $J_n^{\mathbb{K}}$ thus obtained is real and power law associative and thus belongs to the class of what are now known as *Jordan Algebras*.

In fact the list of finite dimensional irreducible homogeneous self-dual cones is quite small and coincides with the list of finite dimensional irreducible Jordan algebras. The list is:

Cone	Algebra	Reduced structure group	Automorphism group
$C(\mathbb{E}^{n-1,1})$	$\Gamma(n-1)$	$SO(n-1,1)$	$SO(n-1)$
$C_n(\mathbb{R})$	$J_k^{\mathbb{R}}$	$PSL(k;\mathbb{R})$	$SO(n)$
$C_n(\mathbb{C})$	$J_j^{\mathbb{C}}$	$PSL(n;\mathbb{C})$	$SU(n)$
$C_n(\mathbb{H})$	$J_k^{\mathbb{H}}$	$SU^\star(2n)$	$Sp(n)$
$C_3(\mathbb{O})$	$J_3^{\mathbb{O}}$	$E_{6(-26)}$	F_4

- $C(\mathbb{E}^{n-1,1}) \subset \Gamma(n-1)$ is the usual Minkowski cone, in $\mathbb{E}^{n-1,1}$ based on a the sphere S^{n-2}. The automorphism group is the Lorentz group $SO(n-1,1)$.

- $C_k(\mathbb{R}) \subset J_n^{\mathbb{R}}$: the set of positive semi-definite $n \times n$ real symmetric matrices. The reduced structure groups is $PSL(n, \mathbb{R})$ and the automorphism group is $SO(n-1)$.
- $C_n(\mathbb{C}) \subset J_n^{\mathbb{C}}$: the set of positive semi-definite $n \times n$ hermitian matrices. The reduced structure group is $PSL(k, \mathbb{C})$ and the automorphism group is $SO(n)$.
- $C_n(\mathbb{H}) \subset J_n^{\mathbb{H}}$: $n \times n$ positive definite quaternionic hermitian matrices. The reduced structure group is $SU^\star(2k)$ and the automorphism group is $Sp(k)$.
- $C_3(\mathbb{O}) \subset J_3^{\mathbb{O}}$: the set of positive semi-definite 3×3 octonionic hermitian matrices. reduced structure group is is $E_{6(-26)}$ and the automorphism group is F_4.

In all cases the automorphism group $Aut(J)$ of the Jordan algebra J is the stability group of the unit element in the algebra, which may be taken as a unit matrix. The reduced structure group of the algebra $Sto_0(J) = PLSG$ is the subgroup of the structure group $G = Str(J)$, leaving the norm of the Jordan algebra invariant.

Note that these results subsume the foundational Alexandrow-Zeeman [56, 57] theorem which states that the auto-morphism group of the causal structure of Minkowski spacetime (defined by the cone $C(\mathbb{E}^{n-1,1})$) consists of dilations and Lorentz transformations [89].

In the case of $\Gamma(n-1)$ one may think of v as an element of the Clifford algebra Cliff$(n-1, 1; \mathbb{R})$, on sets $v = v^\mu \gamma_\mu$. However the Jordan algebra $\Gamma(n-1)$ is generated by γ_i and the identity matrix. Then in all cases the commutative but not associative Jordan product is given by one half the anti-commutator, $u \bullet v = \frac{1}{2}(uv + vu)$. The cone $C(J)$ is then obtained by taking the exponential exp(v) of elements $v \in J$. This is well defined because of the power associativity property $v \bullet v^r = v^{r+1}$ of the algebra.

2.4 Special Cases: Low Order Isomorphisms

It is a striking fact that in the case of 2×2 matrices over $\mathbb{K} = \mathbb{R}, \mathbb{C}, \mathbb{H}$ the Jordan algebras coincide with the Clifford algebras, fact perhaps more familiar in the form

$$Spin(2,1) \equiv SL(2, \mathbb{R}), \tag{23}$$

$$Spin(3,1) \equiv SL(2, \mathbb{C}), \tag{24}$$

$$Spin(5,1) \equiv SL(2, \mathbb{H}). \tag{25}$$

There is also a closely related statement over the octonions for $Spin(9, 1)$ which crops up in string theory.

The middle isomorphism in (25) has lead Penrose to attempt, in his Twistor theory, to connect the use of the complex numbers in spinor analysis with that in quantum mechanics. A connection which moreover seems to give a privileged position to four spacetime dimensions. I think one can take a very different view [59] but to appreciate it we need to make an excursion into

3 Spacetime Signature and the Real Numbers

The basic point being made here is that in 4 + 1, and indeed 9 + 1 and 10 + 1, spacetime dimensions, it is possible, by choosing the spacetime signature appropriately, to develop spinor analysis *at the classical level entirely over the reals*. That is, to consistently use Majorana spinors whose components really are real. In four spacetime dimensions this requires the mainly plus signature convention (the opposite to that which Penrose uses). The complex numbers need only enter when one quantizes.

To see this in more detail we need some facts about Clifford Algebras.

3.1 Clifford Algebras

Given a vector space V[8] with metric g, of signature (s,t) where s counts the positive and t the negative signs, Clifford algebra Cliff$(s,t;\mathbb{R})$ is by definition the associative algebra over the *reals* generated by the relations

$$\gamma_\mu\gamma_\nu + \gamma_\nu\gamma_\mu = 2g_{\mu\nu}, \tag{26}$$

where γ is a basis for V. As a real algebra, the signature does make a difference.

For example

$$\text{Cliff}(0,1;\mathbb{R}) \equiv \mathbb{C}, \tag{27}$$

while

$$\text{Cliff}(0,1;\mathbb{R}) \equiv \mathbb{R}\oplus\mathbb{R}. \tag{28}$$

In fact Cliff$(0,1;\mathbb{R})$ is identical with what are often called 'double numbers' or 'hyperbolic numbers', i.e numbers of the form.

$$a+eb, \qquad a,b\in\mathbb{R}, \qquad e^2=1. \tag{29}$$

As an algebra, Cliff$(0,1;\mathbb{R})$ is not simple, $P_\pm = \frac{1}{2}(1\pm e)$ are projectors onto two commuting sub-algebras.

In a matrix representation

$$i = \begin{pmatrix} 0 & -1 \\ 1 & 0 \end{pmatrix}, \qquad e = \begin{pmatrix} 0 & 1 \\ 1 & 0 \end{pmatrix}. \tag{30}$$

[8] V is *not* $\mathcal{H}_{\text{qm}}$ thought of as real! A good reference for the properties of Clifford algebras used here is [112], see also [111].

However if we pass to the complex Clifford over $\mathbb{C}$ we lose the distinction since

$$\mathrm{Cliff}(0,1;\mathbb{C}) \equiv \mathrm{Cliff}(0,1;\mathbb{C}) \equiv M_2(\mathbb{C}), \tag{31}$$

where $M_2(\mathbb{C})$ is the algebra of all complex valued two by two matrices.

It is precisely at this point that the precautionary principle comes in. We should not rush into adopting

$$\mathrm{Cliff}(3,1;\mathbb{C}) \equiv \mathrm{Cliff}(1,3;\mathbb{C}) \equiv M_4(\mathbb{C}), \tag{32}$$

but rather enquire what are the possible differences between the two signatures[9]. In fact

$$\mathrm{Cliff}(3,1;\mathbb{R}) \equiv M_4(\mathbb{R}), \qquad \mathrm{Cliff}(1,3;\mathbb{R}) \equiv M_2(\mathbb{H}), \tag{33}$$

where

$$\mathbb{H} \equiv \mathrm{Cliff}(0,2;\mathbb{R}) \tag{34}$$

are the quaternions. Despite the differences the spin groups are identical

$$\mathrm{Spin}(3,1) \equiv \mathrm{Spin}(1,3) \equiv SL(2,\mathbb{C}), \tag{35}$$

but if discrete symmetries are taken into account they differ:

$$\mathrm{Pin}(3,1) \neq \mathrm{Pin}(1,3). \tag{36}$$

This has important consequences in spacetimes which are time, space or spacetime non-orientable [13, 32, 35, 36, 38].

3.2 Chiral Rotations

Independently of signature

$$\gamma_5 = \gamma_0\gamma_1\gamma_2\gamma_3, \qquad \gamma_5^2 = -1. \tag{37}$$

Moreover if γ_μ generate a Clifford algebra, then so do

$$e^{\alpha\gamma_5}\gamma_\mu e^{-\alpha\gamma_5} = \cos 2\alpha\gamma_\mu + \sin 2\alpha\gamma_5\gamma_\mu, \qquad \alpha \in \mathbb{R}. \tag{38}$$

Thus by choosing $\alpha = \pi$ we can reverse the sign of the γ_μ and so we expect that no physical consequences should follow from the choice of sign.

[9] A similar point has been made recently by Schucking [62] but he plumps for the quaternions.

The chiral rotations maintain the reality properties of the gamma matrices. Multiplication by i of course reverses the signature.

3.3 Majorana Spinors

It is a striking and, I believe, a possibly rather significant fact that the signature $(3,1)$ leads directly to a Majorana representation, in which all γ matrices are real. Certainly if one holds that $N = 1$ supersymmetry and $N = 1$ supergravity are important, this fact renders the mainly positive signature rather attractive. The precautionary principle would lead one to adopt the signature $(3,1)$ and use a real notation for as long as one can, certainly at the classical level where one need never introduce complex numbers. Thus the basic entities are Majorana spinors ψ belonging to a four dimensional real vector space $\mathbb{M}$ with real, or real Grassmann number components ψ^a, $a = 1,2,3,4$.

Note that if ψ is a Majorana spinor then so is its chiral rotation $e^{\alpha\gamma_5}\psi$.

The charge conjugation matrix $C = -C^t$ satisfies

$$C\gamma_\mu C^{-1} = -\gamma_\mu^t, \qquad C\gamma_5 C^{-1} = -\gamma_5^t. \tag{39}$$

It serves as a Lorentz-invariant symplectic form on $\mathbb{M}$. Thus $\mathrm{Spin}(3,1) \subset \mathrm{Sp}(4;\mathbb{R}) \equiv \mathrm{Spin}(3,2)$.

3.4 Dirac Spinors

To incorporate Dirac spinors, one considers pairs of Majorana spinors ψ^i, $i = 1,2$ which are elements of $\mathbb{R}^4 \oplus \mathbb{R}^4 \equiv \mathbb{R}^4 \otimes \mathbb{C}^2 \equiv \mathbb{R}^8$ If δ_{ij} is the metric and $\epsilon_{ij} = \delta_{ik} J^k{}_j$, the symplectic and $J^k{}_j$ the complex structure which rotates the two summands into each other, we can endow $\mathbb{D} \equiv \mathbb{R}^8$ with a symplectic form ω and a pseudo-riemannian metric g, and hence a pseudo-hermitean structure. In components, for commuting spinors,

$$g(X,Y) = X^{ia} C_{ab} \epsilon_{ij} Y^{ja} = g(Y,X) \tag{40}$$

$$\omega(X,Y) = X^{ia} C_{ab} \delta_{ij} Y^{ja} = -\omega(Y,X), \tag{41}$$

so that

$$\omega(X,Y) = g(JX,Y). \tag{42}$$

The signature of the metric g is $(4,4)$ and of the hermitian form, which is usually written

$$\overline{\psi}\psi, \tag{43}$$

where the Dirac conjugate

$$\overline{\psi} = \psi^\dagger \beta \tag{44}$$

is $(2, 2)$. The 'light cone' on which $\overline{\psi}\psi$ consists of Majorana Spinors

Not that electromagnetic rotations and chiral rotations commute with one another.

Alternatively we can think of the Dirac spinors as elements of a four dimensional complex vector space $\mathbb{D} = \mathbb{M}_{\mathbb{C}} \equiv \mathbb{C}^4$, the complexification of the real space of of Majorana spinors $\mathbb{M}$.

3.5 Weyl Spinors

To see where Weyl spinors fit in we observe that γ_5 acts as a complex structure converting $\mathbb{M} \equiv \mathbb{R}^4$ to $\mathbb{W} \equiv \mathbb{C}^2$. In other words, we write

$$\mathbb{M} \otimes_{\mathbb{R}} \mathbb{C} = \mathbb{D} = \mathbb{W} \oplus \overline{\mathbb{W}}, \tag{45}$$

Elements of W^2 are chiral spinors for which

$$\gamma_5 \psi_R = i \psi_R, \tag{46}$$

Elements of $\overline{W}$ are anti-chiral spinors for which

$$\gamma_5 \psi_L = -i \psi_L, \tag{47}$$

The projectors $\frac{1}{2}(1 - i\gamma_5)$ and $\frac{1}{2}(1 + i\gamma_5)$ project onto chiral and anti-chiral Weyl spinors respectively.

It is of course possible to treat Weyl spinors without the explicit introduction of complex numbers at the expense of introducing pairs of Majorana spinors ψ_1, ψ_2 subject to the constraint that

$$\gamma_5 \psi_1 = -\psi_2, \qquad \gamma_5 \psi_2 = \psi_1. \tag{48}$$

One then has

$$\psi_R = \psi_1 + i\psi_2, \psi_L = \psi_1 - i\psi_2. \tag{49}$$

3.6 Signature Reversal Non-invariance

Many people would argue that after all, a choice of signature is only a convention. That is true, but as we have seen above, this choice of convention comes with consequences. Moreover reversal of signature is not a symmetry of the basic equations of physics. as has emerged very clearly recently in work aimed at understanding why the observed cosmological constant is so small in comparison with its expected value. There have been a number of suggestions [30, 31, 39] that this might be due to a symmetry, analogous to chiral symmetry which is used to account for the smallness of of the electron mass. One candidate for such a symmetry, which may be expressed in a manifestly generally covariant, and simple fashion is the symmetry under change of spacetime signature

$$g_{\mu\nu} \to -g_{\mu\nu}. \tag{50}$$

.

For flat spacetime this is equivalent to the transformation [30]

$$x^\mu \to i x^\mu, \tag{51}$$

taking West Coast to East Coast,

$$\mathbb{E}^{3,1} \to \mathbb{E}^{1,3}, \tag{52}$$

but complexifying or analytic continuation of coordinates are not without problems in curved spacetime and so I prefer (50) which does the job just as well. I have a similar prejudice against formulations in terms of non-generally covariant concepts such as energy [31].

Under (50) one has

$$R_{\mu\nu} \to R_{\mu\nu} \tag{53}$$

and so, if $\Lambda \neq 0$, (50) is definitely not a symmetry of the equations

$$R_{\mu\nu} = \Lambda g_{\mu\nu}. \tag{54}$$

and hence is violated by a non-vanishing cosmological constant.

If scalar fields are present, then (50) is violated by mass or potential terms since under (50) the Christoffel symbols and hence the connection are unchanged

$$\{_\mu{}^\nu{}_\sigma\} \to \{_\mu{}^\nu{}_\sigma\}, \qquad \nabla_\mu \to \nabla_\mu, \tag{55}$$

but the equation

$$g^{\mu\nu}\nabla_\mu\nabla_\nu\phi = V'(\phi) \tag{56}$$

is not invariant and neither is the Einstein equation

$$\frac{1}{8\pi G} R_{\mu\nu} = \partial_\mu\phi\partial_\nu\phi + g_{\mu\nu}V(\phi), \tag{57}$$

On the other hand, the source free Maxwell equations are invariant, but the Einstein equation.

$$\frac{1}{8\pi G}R_{\mu\nu} = g^{\sigma\tau}F_{\mu\sigma}F_{\mu\tau} - \frac{1}{4}g_{\mu\nu}g^{\alpha\beta}g^{\sigma\tau}F_{\alpha\sigma}F_{\beta\tau} \tag{58}$$

is not.

This is part of a more general pattern, for a massless p-form field strength in n spacetime dimensions (so that $p = 1$ corresponds to a scalar and in four dimensions, $p = 3$ to a pseudoscalar or axion) then while the equation of motion is invariant, (50) induces

$$T_{\mu\nu} \rightarrow (-1)^{p+1}T_{\mu\nu}. \tag{59}$$

From this it is clear that the Maxwell equations coupled to a complex scalar field, that is the Abelian Higgs or Landau Ginzburg model are not invariant. This can be seen from

$$\nabla_\nu F^{\mu\nu} = J^\mu. \tag{60}$$

Under (50)

$$F^{\mu\nu} \rightarrow F^{\mu\nu}, \tag{61}$$

but

$$J^\mu \rightarrow -J^\mu. \tag{62}$$

Similarly the Lorentz equation

$$\frac{d^2x^\mu}{d\tau^2} + \{_\sigma{}^\mu{}_\tau\}\frac{dx^\sigma}{d\tau}\frac{dx^\tau}{d\tau} = \frac{e}{m}g^{\mu\alpha}F_{\alpha\beta}\frac{dx^\beta}{d\tau}, \tag{63}$$

is not invariant under (50).

Thus the classical equations of motion of the bosonic sector of the standard model are certainly not invariant under (50). To make them so would entail adding additional fields whose energy momentum tensor is opposite in sign to the standard case. These fields would antigravitate rather than gravitate. Various schemes of this sort have been discussed in the literature (e.g. [41–43]).

If, therefore, the signature reversal is not a symmetry of our world, then it seems reasonable to me to suppose that one signature is preferred over the other, and that is the view being advocated here.

Of course one could follow Duff and Kalkkinen that we have simply mistaken the dimension we are in, [64, 65] or conclude that the signature of spacetime may vary from place to place, some regions having signature $-+++$ and some signature $+---$ [63]. Perhaps one should say that spacetime signature is an emergent property.

4 More Than One Time: Signature Change

We have been arguing that time, or at least a universal complex structure on the quantum mechanical Hilbert space single may be an emergent, or historical phenomenon.

This seems clearest in certain, instanton based, approximate, treatments of the birth of the universe based on what Hartle and I have called Tunnelling Metrics [15], in which a Riemannian manifold M_R and and a Lorentzian manifold M_L are joined on across a surface Σ of time symmetry which may be regarded as the origin of time surface. There is no time in M_R where the metric signature is $++++$. The metric signature flips to $-+++$ across Σ. If that can happen why can't it flip to $--++$ across some other surface, as suggested by Eddington long ago [85]?

Signature flip also arises in brane-world scenarios in which the brane bends over in time while remaining a smooth sub-manifold of the Lorentzian bulk spacetime,ceases to be *timelike*, but rather spacelike with positive definite (i.e. Riemannian) induced metric [86–88]. However unless the bulk as more than one time the transition can only be from Lorentzian to Riemannian. In the model studied in [86] time certainly emerges after the collision of two branes.

The question therefore arises, could two, or possibly more than two times have emerged? There has been a fairly large amount of work on the possibility of two or more times, i.e on spacetimes of signature $--,+,+,+$ or $-,-,-,+,+,+$. A very early example is hinted at by Halsted [113] A later, and for me difficult to understand example is [93], where the extra temporal coordinate is called 'anti-time 'or 'eternity').

As far as a can see, little attempt to relate them to the algebraic structure of quantum mechanics, although a theory of Kostant comes quite close.

In fact, the standard reason for rejecting such theories is the existence of the instabilities and causality violations that result as a consequence of the fact that the interior of the light cone is no longer convex. This is clearly shown by Dorling's argument [45] that the lowest mass particle is such a spacetime could decay into particles of heavier mass. In Kaluza–Klein theories, timelike extra dimensions lead to negative energies for vector fields on dimensional reduction [46] and provides limits on their size [90].

One way to say this is that necessarily such spacetimes cannot admit a *time orientation* and hence, in accordance with our general outlook, cannot admit standard complex quantum mechanics.

Among multi-time theories, a particularly intriguing case from the mathematical point of view is that of six-dimensional manifolds with neutral or Kleinian signature $(+++---$. In other words where there is a complete symmetry between space and time. This has been energetically pursued by Cole over many years [95–105] in an attempt to make physical sense of it. I am skeptical but believe it may ultimately play a role in string theory.

One has the isomorphisms

$$SO(3,3) \equiv SL(4,\mathbb{R})/\mathbb{Z}_2, \qquad \mathrm{Cliff}(3,3;\mathbb{R}) \equiv M_8(\mathbb{R}). \tag{64}$$

The first isomorphism links us to real three-dimensional projective geometry, via a a real form of Twistor theory [59]. One may think of $\mathbb{E}^{3,3}$ as the space of bi-vectors in $L^{\mu\nu} = -L^{\nu\mu}$ in $\mathbb{R}^4$ endowed with the metric

$$\frac{1}{4}\epsilon_{\mu\nu\sigma\tau}L^{\mu\nu}L^{\sigma\tau}. \tag{65}$$

By the well-known Plücker correspondence, lines in $\mathbb{RP}^3$ correspond to simple bi-vectors in and hence to to null six-vectors in $\mathbb{E}^{3,3}$. It is also possible to regard $\mathbb{R}^4$ or its projectivization $\mathbb{RP}^3$ as the space of Majorana spinors in four spacetime dimensions. Conformally $SO(3,3)$ is the conformal group of $\mathbb{E}^{2,2}$. In Penroses's Twistor Theory one complexifies and another real form is $SO(4,2)$ the conformal group of ordinary Minkowski spacetime $\mathbb{E}^{3,1}$.

Kostant [106, 107] has made the imaginative proposal that our spacetime (with signature $(3,1)$) is a three-brane embedded in a six-dimensional bulk spacetime with a metric of signature $(3,3)$. The restriction of the ambient metric to the normal bundle has signature $(0,2)$ and the associated $SO(2)$ symmetry allows him to think of the normal bundle as a complex line bundle over spacetime. This is the origin of electromagnetism in his theory.

To obtain examples, Kostant noted that the conformal group of $\mathbb{E}^{3,3}$ is $SO(4,4)$. More accurately $SO(4,4)$ acts globally on the conformal compactification of $\mathbb{E}^{3,3}$, which may be regarded as the space of null rays in $\mathbb{E}^{4,4}$. To get compactified Minkowski spacetime ($S^1 \times S^3/\mathbb{Z}_2$ one intersects the null cone of the origin of $\mathbb{E}^{4,4}$ with a 6-plane through the origin of signature $(4,2)$. An interesting aspect is that since we are dealing with $SO(4,4)$ there is a triality which acts. Mathematically, Kostant's model has many intriguing features (see [108, 109]) but so far clearly fails to make much contact with the real world.

Another, purely technical use of three times is to study integrability. Because $\mathbb{E}^{3,3}$ admits a para-hermitean structure, in other words it admits an isometric involution J on the tangent space which, $J^2 = 1$, (i.e. para-complex) and such that $g(JX, JY) = g(X, Y)$, this may be used to obtain the KP equations via a self-duality condition on $\mathfrak{sl}(2, \mathbb{R})$ gauge fields [110].

5 Examples

The general algebraic considerations considerations may seem rather abstract, but they have already arisen in the application of quantum mechanics to cosmology.

In what follows, I shall give some examples. Before doing so I note that

The much discussed question of whether black hole evaporation is unitary is **meaningless** *if there is no complex structure, and ill-posed if there is more than one.*

5.1 Quantum Field Theory in Curved Spacetime

In Quantum Field Theory in Curved Spacetime the main problem is that there is no unique definition of "positive frequency". In the free theory, $V = \mathcal{H}_{\text{one particle}}$ is the space of real-valued solutions of wave equations. V is naturally (and covariantly) a symplectic (boson), or orthogonal (fermion)[10] vector space

$$\omega(f,g) = \int (\dot{f} g - \dot{g} g) d^3 x = -\omega(g,f) \tag{66}$$

$$g(\psi,\chi) = \int (\psi^t \chi) d^3 x = g(\chi,\psi) \tag{67}$$

To quantize we complexify and decompose

$$V_{\mathbb{C}} = \mathbb{C} \otimes V = V^+ \oplus V^- \tag{68}$$

This decomposition (which defines a complex structure) [10,34,115] is not unique.

This non-uniqueness corresponds physically to the possibility of particle production and is an essential part of our current understanding of **black hole evaporation** *and* **inflationary perturbations**.

At this point it may be instructive to recall [117] why commutation relations of the form

$$\left[\hat{x}^\mu, \hat{P}_\nu\right] = i\delta^\mu_\nu \tag{69}$$

don't apply in quantum field theory in Minkowski spacetime. If they did, then they would have, up to natural equivalence, to be represented in the standard Stone–Von-Neumann fashion on $L^2(\mathbb{E}^{3,1})$. But then the energy $\hat{P}^0$, could not be bounded below. Thus $L^2(\mathbb{E}^{3,1})$ is not the quantum mechanical Hilbert space. Rather, as stated above, it is the space of positive frequency solutions of the Klein–Gordon or Dirac equations. These are is much more subtle objects and certainly not uniquely defined in a curved spacetime manifold $\{\mathcal{M}, g\}$, unlike $L^2(M, \sqrt{-g} d^4 x)$, the obvious generalization of $L^2(\mathbb{E}^{3,1})$, which is unambiguous even in a curved spacetime.

5.2 The Wave Function of the Universe

In *Hartle and Hawking's Wave Function for The Universe*

$$\Psi(h_{ij}, \Sigma) = \int d[g] e^{-I_{\text{euc}}(g)}, \quad h_{ij} = g_{ij}|_{\Sigma=\partial M} \tag{70}$$

Is real valued. To get a notion of time one typically passes to a Lorentzian WKB approximation S_c

$$\Psi = A e^{iS_c} + \bar{A} e^{-iS_c} \tag{71}$$

but this is only a semi-classical approximation, in other words

[10]We use real (Majorana) commuting spinors for convenience: there use is not essential.

Time, the complex numbers and the complex structure i_{qm} of quantum mechanics **emerge** *only as an approximation at late times.*

5.3 Euclidean Quantum Field Theory

In fact in Euclidean Quantum Field Theory it is not sufficient just to compute correlators.

In order to recover*Quantum Mechanics*, rather than merely to indulge in an unphysical case of *Statistical Mechanics*, the correlators must exhibit *Reflection Positivity* [20, 21]. This guarantees the possibility of analytically continuing to real time.

This can be done for Riemannian backgrounds if they admit a suitable reflection map, for example static or time-symmetric metrics such as Real tunneling geometries [9, 15–19, 22–24].

However most Riemannian metrics do not admit such a reflection map.

Thus **generically** *in such approaches one would not recover standard complex quantum mechanics. Only for very special classical saddle points of the functional integral would a well defined complex structure emerge.*

5.4 Lorentzian Creation ex nihilo

The next example involves a Lorentzian Born From Nothing Scenario [13, 14, 116]. Essentially, one considers de-Sitter spacetime modded out by the antipodal map $dS/\mathbb{Z}_2$ (so-called elliptic interpretation).

$$-(X^0)^2 + (X^1)^2 + (X^2)^2 + (X^3)^2 + (X^4)^2 = \frac{3}{\Lambda} \qquad \mathbb{Z}_2 : X^A \equiv -X^A. \quad (72)$$

Now the antipodal map preserves space orientation but reverse time orientation. But in quantum mechanics a time reversing transformation is represented by an anti-unitary operator Θ and if all states are invariant up to a factor

$$\Theta\Psi = \lambda\Psi \quad (73)$$

then *only real linear combinations are allowed.*

Thus *Quantum Mechanics in $dS/\mathbb{Z}_2$ is Real Quantum Mechanics.*

This jibes with the fact that under the action of the antipodal map is *antisymplectic* on the bosonic space of solutions V

$$\omega(\cdot,\cdot) \rightarrow -\omega(\cdot,\cdot). \quad (74)$$

This renders imposing the CCR's impossible[11]

Compare regular time reversal

$$(p_i, q^i) \to \omega = (-p_i, q^i) \Longrightarrow dp_i \wedge dq^i \to -dp_i \wedge dq^i = -\omega \qquad (75)$$

If there is no symplectic form then the Heisenberg commutation relations make no sense, one cannot geometrically quantize.

This pathology arises quite generally for spacetimes which do not admit a *Time Orientation*, i.e. a smooth choice of future lightcone. Such spacetimes always have a double cover which is time orientable and so may be regarded as the quotient of a time orientable spacetime by a generalized , time orientation reversing, antipodal map. The double cover thus realizes various speculative ideas of the past and not so recent past [47–51] of spacetimes in which the arrow of time runs one way in one part and the other way in the other.

In other words quantum field theory is not defined unless one may define an *Arrow of Time*[12].

Amusingly CTC's seem to be quiet innocuous from this point of view. It seems that they can be compatible with quantum mechanics, but not necessarily locality.

6 Topology, Time Reversal and the Arrow of Time

An interesting question, discussed by Chamblin and myself [26], is whether this arrow is intrinsically defined, or whether both possibilities are on the same footing.

In other words, do there exist time-orientable spacetimes which have an *intrinsic direction of time?*

The analogy here is with a quartz crystal which is either left-handed or right handed. This is because the point group contains no reflections or inversions.

For a spatial manifold Σ one asks: does Σ there exist an orientation reversing diffeomorphism. In other words is there a diffeomorphism taking Σ with one orientation to Σ with the opposite orientation?. For such manifolds a *Parity Map* cannot be defined. Such "handed" manifolds are quite common, certain Lens Spaces and $\mathbb{CP}^2$ being examples[13].

For spacetimes the analogous question is whether there exist a time reversing diffeo Θ?

We found some rather exotic examples, based on higher dimensional Taub-NUT spacetimes for which no such diffeo Θ exists.

---

[11]Bernard Kay has implemented this argument more rigourously within an algebraic framework [120].

[12]Amusingly CTC's seem to be quiet innocuous from this point of view. It seems that they can be compatible with quantum mechanics.

[13]See Hartle and Witt [114].

The question can be formulated in *Hamiltonian Mechanics*. Does there exist a symplectic manifold $\{M, \omega\}$ admitting no anti-symplecto-morphism, i.e. a time reversal map Θ such that

$$\Theta^\star \omega = -\omega. \tag{76}$$

The answer to this topological question, which should not be confused with asking whether any particular Hamiltonian, on a symplectic manifold which *does* admit a time reversal map, is time-reversal invariant, i.e. whether

$$\Theta^\star H = H\,?, \tag{77}$$

does not seem to be known.

7 Unification and Spin(10)

If the viewpoint advocated here is on the right track, one might expect that should be signs in what little information we have about possible unification schemes. A very popular one is based on the group $SO(10)$ and it is perhaps gratifying that it seems to fit with the philosophy espoused here.

In the standard electro-weak model, the neutrinos are purely left-handed and a description of the fundamental degrees of freedom in terms of Weyl spinors is often felt to be appropriate. One may then argue that this more more convenient with the mainly minus signature. However nothing prevents one describing it using Majorana notation and the mainly plus signature. Moreover the discovery of the non-zero neutrino masses and the so-called see-saw mechanism make it plausible that there is a right handed partner for the neutrinos and the fact that then each family would fit into a chiral (i.e. **16**) representation of Spin(10) makes it perhaps more attractive to describe the fundamental fields in Majorana notation. This would tend to favour the use of the mainly plus signature.

To see this in more detail recall[14]

$$\text{Cliff}(10, 0; \mathbb{R}) \equiv M_{32}(\mathbb{R}). \tag{78}$$

Let Γ_a, $a = 1, 2, \ldots, 10$ be a representation of the generators by real 32×32 matrices and

$$\Gamma_{11} = \Gamma_1 \Gamma_2 \ldots \Gamma_{10}, \tag{79}$$

so that[15]

$$\Gamma_{11}^2 = -1. \tag{80}$$

---

[14]This is clear from the periodicity modulo eight of Clifford algebras $\text{Cliff}(s+8, t) \equiv \text{Cliff}(s, t) \otimes M_{16}(\mathbb{R})$ and the easily verified fact that the that $\text{Cliff}(2, 0; \mathbb{R}) \equiv M_2(\mathbb{R})$.

[15]The matrices Γ_a, Γ_{11} generate the M-theory Clifford algebra $\text{Cliff}(10, 1; \mathbb{R}) \equiv M_{32}(\mathbb{R}) \oplus M_{32}(\mathbb{R})$.

It is customary to describe the Spin(10) model in terms of 16 left handed spacetime Weyl fermions which are then placed in a single complex chiral **16**, Ψ of Spin(10)

$$\Gamma_{11}\Psi = i\Psi, \tag{81}$$

but this is completely equivalent ,and notationally simpler to regard the 16 spacetime Weyl fermions as 32 spacetime Majorana fermions and then to regard Ψ as a 32 dimensional Majorana spinor of Spin(10) subject to the constraint

$$\Gamma_{11}\Psi = \gamma_5\Psi. \tag{82}$$

In more detail, we start with the 15 observed left handed Weyl fermions of the electro-weak theory with their weak hypercharges $Y = Q - t_3$, where Q is the electric charge and t_3 the third component of weak iso-spin

$$\begin{pmatrix} u_L \\ d_L \end{pmatrix}, \; Y = \frac{1}{6} \qquad \begin{pmatrix} \nu_L \\ e_L \end{pmatrix}, Y = -\frac{1}{2} \tag{83}$$

$$u^c_L, \; Y = -\frac{2}{3} \qquad d^c_L, \; Y = \frac{1}{3} \qquad e^c_L, Y = 1. \tag{84}$$

The first row consists of four iso-doublets and the second row of seven iso-singlets. The up and down quarks u_L and d_L are in a **3** of $SU(3)$ colour and their charge conjugates u^c_L, d^c_L are in a $\bar{\mathbf{3}}$ of $SU(3)$. In fact the, because effective group is $S(U(3) \times U(2)) \equiv SU(3) \times SU(2) \times U(1)/\mathbb{Z}_3 \times \mathbb{Z}_2$, where $\mathbb{Z}_3$ and $\mathbb{Z}_2$ are the centres of $SU(3)$ and $SU(2)$ respectively [33]. This is because the electric charge assignments are such that acting with $\mathbb{Z}_3 \times \mathbb{Z}_2 \equiv \mathbb{Z}_6$ can always be compensated by a $U(1)$ rotation.

Now $S(U(3) \times U(2))$ is a subgroup of $SU(5)$ and is well known one may fit all 15 left handed Weyl spinors in a **5** and a **10.** However it is more elegant to adjoin the charge conjugate of the right handed neutrino, ν^c_L to make up a complex **10** of Spin(10). In fact the multiplets may be organized into multiplets of the Spin(6) $\times$ Spin(4) $\equiv SU(4) \times SU(2) \times SU(2)$ subgroup of Spin(10)

$$\begin{pmatrix} u_L \\ d_L \end{pmatrix}, \qquad \begin{pmatrix} \nu_L \\ e_L \end{pmatrix}. \tag{85}$$

$$\begin{pmatrix} u^c_L \\ d^c_L \end{pmatrix}, \qquad \begin{pmatrix} \nu^c_L \\ e^c_L \end{pmatrix}. \tag{86}$$

In this formalism we have left-right symmetry with the first row consisting of four weak iso-doublets and the bottom row of four doublets of some other, as yet unobserved $SU(2)$. The quarks and leptons also form two Spin(6) $\equiv SU(4)$ quartets.

8 Gravitational CP Violation?

To conclude I would like to illustrate once more the advantages of the reality viewpoint by addressing a question of some current interest which is relevant to the present proceedings. That is whether CP-violating Dirac and Majorana mass terms for spin half fermions can give rise to detectably different behaviour as the particles fall in a gravitational field of a rotating body, due to the Lense–Thirring effect [66–69].

If they could, then a violation of the Weak Equivalence Principle in the form of the Universality of Free Fall would be entailed, which seems rather unlikely. The calculations given in [66, 67] are rather complicated and in view of the great importance of the issue, it seems worth while examining the question in a more elementary fashion. There are also potential implications for the quantum theory of black holes.

There are two aspects of the problem:

- The emission and detection of the fermions by ordinary matter
- Their propagation from source to detector through an intervening gravitational field

It is the latter which I will be discuss here If the fermions are assumed to be electrically neutral and with vanishing electric and magnetic dipole moments, this is a well defined problem in general relativity. Clearly if the fermions are moving in an electromagnetic field and they possess electrc charges and/or magnetic and electric dipole moments the conclusions might be modified, but then the question is no longer one of pure gravity.

With our conventions, A system of k Majorana fermions ψ has Lagrangian

$$L = \frac{1}{2}\psi^t C \not{D} \psi - \frac{1}{2}\psi^t C \big(m_1 + m_2 \gamma^5\big)\psi. \tag{87}$$

where m_1 and m_2 are real symmetric $k \times k$ matrices.

The kinetic term, but not the mass term, is invariant under $SO(k)$ transformations

$$\psi \rightarrow O\psi, \qquad O^t O = 1. \tag{88}$$

Note that one may write

$$O = \exp \omega_{ij}, \qquad \omega_{ij} = -\omega_{jk}. \tag{89}$$

The kinetic term, but not the mass term is also invariant under chiral rotations

$$\psi \rightarrow P\psi, \tag{90}$$

$$P = \exp \nu_{ij} \gamma^5, \qquad \nu_{ij} = \nu_{ji} \tag{91}$$

Combining these two sets of transformations we see that the kinetic term, but not the mass term is in fact invariant under the action of $U(k)$, i.e. under

$$\psi \to S\psi, \tag{92}$$

$$S = \exp\left(\omega_{ij} + \nu_{ij}\gamma^5\right). \tag{93}$$

The $U(k)$ invariance is perhaps more obvious in a Weyl basis. Since

$$\left(\gamma^5\right)^2 = -1, \tag{94}$$

one may regard γ^5 as providing a complex structure on the space of $4k$ real dimensional Majorana spinors, converting it to the $2k$ complex dimensional space of positive chirality Weyl spinors for which

$$\gamma^5 = i. \tag{95}$$

Clearly S then becomes the exponential of the $k \times k$ anti-hermitian matrix

$$\omega_{ij} + i\nu_{ij}. \tag{96}$$

Thus

$$SS^\dagger = 1. \tag{97}$$

The mass matrix is then a complex symmetric matrix

$$m = m_1 + im_2, \tag{98}$$

and under a $U(k)$ transformation

$$m \to S^t m S. \tag{99}$$

At this point we invoke the result of Zumino [70] that S may chosen to render the matrix m diagonal with real non-negative entries.

This implies that k free massive Majorana (or equivalently Weyl) fermions ψ^i moving in a gravitational field will satisfy

$$\not{D}\psi^k - \mu_k\psi^k = 0,$$

with no sum over k and where the masses μ_k may be taken to be real and non-negative. There are no exotic non-trivial effects moving past a spinning object due to the Lense–Thirring effect. In particular there are no CP violating effects and gravity alone cannot distinguish ‘Majorana’ from ‘Dirac’ masses.

8.1 Behaviour in a Gravitational Field

From now on, we assume that the mass matrix m is real and diagonal. If one iterates the Dirac equation and uses the cyclic Bianchi identity in a curved space one gets

$$-\nabla^2\psi + \frac{1}{4}R\psi + m^2\psi = 0. \tag{100}$$

As is well known, there is no 'gyro-magnetic' coupling between the spin and the Ricci or Riemann tensors [71]. To proceed, one may pass to a Liouville–Green-Wentzel–Kramers–Brilouin approximation of the form

$$\psi = \chi e^{iS}. \tag{101}$$

One obtains

$$(i\gamma^\mu \partial_\mu S + m)\chi = 0, \tag{102}$$

and

$$\partial_\mu S \nabla^\mu \chi = 0. \tag{103}$$

The analogue of the Hamilton Jacobi equation is

$$\left(g^{\mu\nu}\partial_\mu S \partial_\nu S + m^2\right)\chi = 0. \tag{104}$$

Now since m is diagonal with diagonal entries μ_i, say, then each eigenspinor χ_i propagates independently along timelike geodesics via

$$\mu_i \frac{dx^\mu}{d\tau} = g^{\mu\nu}\partial_\mu S. \tag{105}$$

The spinor amplitude χ_i is parallelly transported along these geodesics. Of course the geodesics are independent of the mass eigenvalue μ_i and the polarization state given by χ_i. Indeed if the fermion starts off in a given polarization state with (with the associated mass), it remains in it. In other words, at the L-G-W-K-B level, the Weak Equivalence Principle, in the form of the Universality of Free Fall holds

9 Pure States ⟶ Mixed States?

The completely thermal character of Hawking radiation (*at the semi-classical level*) and the apparent violation of Global Symmetries if black hole decay leaves no remnants led Hawking [80] to suggest that while the standard propositional structure of quantum mechanics, and its complex structure, should remain in a full quantum theory of gravity, the evolution law should change. In particular the evolution law should allow pure states to evolve to mixed states. In what follows I shall review the formalism suggested (and now abandoned) by Hawking and then comment on its relation to the suggestion I am making about the complex structure of quantum mechanics. I shall also relate this discussion to issues of reversibility and the arrow of time.

9.1 Density Matrices

Are positive semi-definite Hermitean operators acting on a quantum mechanical Hilbert space $\mathcal{H}$ with unit trace

$$\rho = \rho^\dagger, \qquad \mathrm{Tr}\rho = 1, \qquad \langle\psi|\rho\psi\rangle \geq 0, \quad \forall\, |\psi\rangle. \tag{106}$$

If one diagonalizes

$$\rho = \sum_n P_n |n\rangle\langle n| \tag{107}$$

where $P_n \geq 0$ is the probability one is in the (normalized) state $|n\rangle$, and

$$\sum_n P_n = 1. \tag{108}$$

A pure state is one for which

$$\mathrm{Tr}\rho = 1, \tag{109}$$

in which case, all but one of the P_n vanishes and one is in that state with certainty. In a general orthonormal basis with one writes

$$\rho = \rho_{mn} |m\rangle \otimes \langle n| \tag{110}$$

with

$$\rho_{mm} = 1, \qquad \rho_{mn} = \bar{\rho}_{nm} \tag{111}$$

There is a distinguished density matrix ι associated with complete ignorance for which $P_n = \frac{1}{N}$ where $N = \dim_{\mathbb{C}} \mathcal{H}$,

9.2 Gibbs Entropy

Normalized density matrices form a convex cone in the space of all Hermitean operators and the Gibbs entropy

$$S = -\mathrm{Tr}\rho \ln \rho = -\sum_n P_n \ln P_n \tag{112}$$

is a convex function on the cone which is largest at the completely ignorant density matrix ι. and vanishes for any pure state.

If $N = 2$ we may set

$$\rho = \frac{1}{2}\big(x^0 \mathbb{I}_2 + x^i \sigma_i\big) \tag{113}$$

where σ_i, $i = 1, 2, 3$ are Pauli matrices and the cone corresponds to the future light cone of four dimensional Minkowski spacetime

$$x^0 \geq \sqrt{x^i x^i} = r. \tag{114}$$

The unit trace condition implies that $x^0 = 1$ and thus $r - \sqrt{x^i x^i} \leq 1$ One finds that

$$S = -\ln[(\frac{1+r}{2})^{\frac{1+r}{2}}(\frac{1-r}{2})^{\frac{1-r}{2}}]. \tag{115}$$

The entropy is maximum at the origin and goes to zero on the boundary of unit ball.

9.3 Evolution by an S-Matrix

In general we might be interested in situations where there is an in and an out Hilbert space. $\mathcal{H}^{\text{in}}$ and $\mathcal{H}^{\text{out}}$ respectively. Conventionally on thinks of $\mathcal{H}^{\text{in}}$ and $\mathcal{H}^{\text{out}}$ as being isomorphic, except possibly described in a different basis but one could envisage more general situations. One has an associated set of states or density matrices for $\mathcal{H}^{\text{in}}$ and $\mathcal{H}^{\text{out}}$. The set of such (unormalized) mixed states we call $\mathcal{N}^{\text{in}}$ or $\mathcal{N}^{\text{out}}$ respectively.

Conventionally one postulates there is a unitary map $S : \mathcal{H}^{\text{in}} \rightarrow \mathcal{H}^{\text{out}}$ called an S-matrix such that

$$|\text{out}\rangle = S|\text{in}\rangle \tag{116}$$

which acts by conjugation on mixed states or density matrices

$$\rho^{\text{out}} = S\rho^{\text{in}} S^{\dagger}. \tag{117}$$

9.4 Tracing Out

A situation which often arises is when the out Hilbert space$\mathcal{H}^{\text{out}}$ is a tensor product

$$\mathcal{H}^{\text{out}} = \mathcal{H}^{\text{out}\,1} \otimes \mathcal{H}^{\text{out}\,2} \tag{118}$$

An initial state |in⟨ which remains pure will have an expansion

$$|\mathrm{in}\rangle = c_{mM}|m\rangle \otimes |M\rangle \tag{119}$$

where $|m\rangle$ is a basis for $\mathcal{H}^{\mathrm{out}\,1}$ and $|M\rangle$ a basis for $\mathcal{H}^{\mathrm{out}\,2}$. An observable O_1 which acts as the identity on $\mathcal{H}^{\mathrm{out}\,2}$ will have an expectation value

$$\langle \mathrm{in}|O_1|\mathrm{in}\rangle = \rho_{mn}|m\rangle \otimes \langle n|, \tag{120}$$

where

$$\rho_{mn} = \bar{c}_{mM}c_{nM}, \tag{121}$$

where we use the fact that

$$\langle n|O_1|m\rangle = \mathrm{Tr}\big(0_1|m\rangle \otimes \langle n|\big). \tag{122}$$

In other words observations made only in $\mathcal{H}^{\mathrm{out}\,1}$ can tell us nothing about $\mathcal{H}^{\mathrm{out}\,2}$ and hence will in general behave as is the final state were mixed.

9.5 Evolution by an \$ Matrix

Taking $\mathcal{H}^{\mathrm{out}\,1}$ to be states at infinity and $\mathcal{H}^{\mathrm{out}\,2}$ horizon states shows that in general outgoing radiation from a black holes with a permanent horizon will be in a mixed state.

However back reaction means that the horizon is not permanent and the issue arises whether taking back reaction into account would give a pure or a mixed state.

More generally, one may try to construct a generalization of standard quantum mechanics in which in general pure states evolve to mixed states. One postulates that there is a linear map $\$: \mathcal{N}^{\mathrm{in}} \to \mathcal{N}^{\mathrm{in}}$ such that

$$\rho^{\mathrm{out}} = \$\rho^{\mathrm{in}}. \tag{123}$$

One further postulates that \$ is hermiticity, and trace-preserving

$$(a) \qquad (\$\rho)^\dagger = \rho^\dagger, \tag{124}$$

$$(b) \qquad \mathrm{Tr}\$\rho = \mathrm{Tr}\rho, \tag{125}$$

$$(c) \qquad \$\iota = \iota. \tag{126}$$

One also demands that \$ takes positive semi-definite operators to positive semi-definite operators.

Some comments are in order.

- The assumption of linearity, is a form of locality assumption since it amounts to assuming 'non-interference of probabilities'. It should be possible to lump together results of two independent experiments and obtain the same probabilities.

  Thus if in one ensemble consisting of 100 states with 30 in state 1 and 70 in state 2 these go to states 3 and 4 in 45 and 55 times respectively, and in a second run of the same experiment 30 in state 1 and 70 in state 2 go to 72 and 28 in states 3 and 4 respectively than it should be the case, if the usual idea of probabilities is to make sense, that run in which $85 = 30 + 55$ are in state 1 and $115 = 70 + 45$ in state 2, then $117 = 45 + 72$ should land up in state 3 and $83 = 55 + 28$ should land up in state 4.

  Of course strictly speaking, this argument only works for *commuting* density matrices but, by continuity it seems reasonable to assume linearity for all density matrices.
- The assumption that the completely ignorant density matrix ι is preserved in time would seem to be necessary for any type of thermodynamics to be possible, not least because the completely ignorant density matrix ι has the largest Gibbs entropy.

9.6 *Invertibility and Factorisability*

Standard S-matrix evolution is such that

$$\$\rho = S\rho S^{\dagger}. \tag{127}$$

Such \$-matrices are said to be *factorisable* and factorisable density matrices clearly take pure states to pure states, but a general \$ matrix will take pure states to mixed states. In fact, in general, a \$ matrix acts as a contraction on the convex cone of positive definite Hermitian operators. Thus in general it is not invertible [81, 82]. Indeed there is a

Theorem: *A super-scattering matrix \$ is invertible iff it is factorisable*

Proof: Assume the contrary. Then there exits a mixed out-state ρ^{out} which is mapped to a pure state $\$\rho^{\text{out}} = |\text{in}\rangle$. Thus

$$\$\sum_n P_n |n, \text{out}\rangle \otimes \langle n, \text{out}| = |\text{in}\rangle \otimes \langle \text{in}| \tag{128}$$

□

Let $|\psi^{\text{in}}\rangle$ be any in state orthogonal to $|\text{in}\rangle$ One has

$$\sum_n P_n \langle \psi^{rmin}| \Big(\$|n, \text{out}\rangle \langle n, \text{out}| \Big) |\psi^{\text{in}}\rangle = 0. \tag{129}$$

But $\$|n,\mathrm{out}\rangle\langle n,\mathrm{out}|$ is a density matrix and so positive semi-definite. Thus a if $|n,\mathrm{out}\rangle$ has $P_n \neq 0$, then it must be orthogonal to every pure state $|\psi^{\mathrm{in}}\rangle$ orthogonal to $|\mathrm{in}\rangle$ and hence

$$\$|n,\mathrm{out}\rangle = |\mathrm{in}\rangle \otimes \langle \mathrm{in}|, \qquad \forall\,\{P_n|P_n \neq 0\}. \tag{130}$$

But if \$ takes *all* such states $|n,\mathrm{out} >$ to the same state $|\mathrm{in}\rangle$ it cannot be invertible.

9.7 Irreversibility and CPT

Thus, as one might have expected, evolution by a superscattering matrix would irreversible. How does this square with our prejudices about *CPT*? This is usually taken to be an anti-unitary invertible (since $\theta^2 = 1$) $\theta : \mathcal{N}^{\mathrm{out}} \to \mathcal{N}^{\mathrm{out}}$ which takes pure states to pure states, and preserves traces and preserves ignorance.

Let us call its restriction to pure states Strong CPT.

9.8 Strong CPT

assumes an invertible map Θ from in states to out states

$$\Theta = \$\Theta^{-1}\$. \tag{131}$$

Thus

$$\$^1 = \Theta^{-1}\$\Theta^{-1}. \tag{132}$$

In other words Strong CPT implies that the evolution is invertible. Note that this rather strong result does not assume that either Θ or \$ is a linear map. However if \$ satisfies the requirements for a superscattering matrix and strong CPT, then it must be invertible and hence factorisable.

9.9 Weak CTP

Faced with the result above, one could argue that only *probabilities* are related by CPT. this

$$\mathrm{Prob}(|\psi\rangle \to |\phi\rangle) = \mathrm{Prob}(\Theta^{-1}|\phi\rangle \to \Theta|\psi\rangle). \tag{133}$$

That is

$$\langle\phi|\$\Big(|\psi\rangle\langle\psi|\Big)\phi\rangle = \langle\Theta\phi|\$\Big(|\Theta^{-1}\phi\rangle\langle\Theta^{-1}\phi|\Big|\Theta\psi\rangle, \tag{134}$$

that is

$$\$^{\dagger} = \Theta^{-1} \$ \dagger \Theta^{-1}. \tag{135}$$

Of course for a factorisable \$ matrix (135) holds by unitarity of the S matrix.
Moreover (135) implies that the superscattering operator is ignorance preserving

$$\$ i = i. \tag{136}$$

An interesting set of questions is

- Is (135) equivalent to *detailed balance*?
- Does (135) imply the H theorem?
- Does (135) imply that only the microcanonical ensemble, i.e. the perfectly ignorant density matrix i is left-invariant by \$?

A full answer to these questions appears not be known but what is well known is the situation when all density matrices are assumed diagonal.

9.10 Pauli Master Equation

This is essentially the case when the density matrix remains diagonal. One sets

$$\dot{P}_r = \sum_{s \neq r} U_{rs} P_s - P_r \sum_{s \neq r} U_{sr} \tag{137}$$

where $U_{rs} \geq 0$ may be interpreted as the transition probability per unit time if a transition from state $|s\rangle\langle s|$ to state $|r\rangle\langle r|$.
In perturbation theory

$$U_{rs} = |\langle r|H_{\text{pert}}|s\rangle|^2 = \langle r|H_{\text{pert}}|s\rangle^{\star}\langle r|H_{\text{pert}}|s\rangle \tag{138}$$

and hence from the Hermiticity of the Hamiltonian

$$\langle r|H_{\text{pert}}|s\rangle^{\star} = \langle s|H_{\text{pert}}|r\rangle \tag{139}$$

we have *detailed balance* or *microscopic reversibility*

$$U_{rs} = U_{sr} \tag{140}$$

Under this assumption and that all transitions take place, i.e $U_{rs} > 0 \quad \forall\, r, s$ we have the two following [75–79]

Theorem A. *(Existence Uniqueness of Equilibrium) there is a unique equilibrium state* ι *of total ignorance such that* $P_r = P_s, \quad \forall r, s$ and

Theorem B. *('H-Theorem') The entropy* $S = -\sum_r P_r \ln P_r$ *is monotonic increasing* $\dot{S} \geq 0$.

Proof of A. under these assumptions

$$\dot{P}_r = \sum_{s,s\neq r} U_{rs}(P_s - P_r). \tag{141}$$

If we order the P_s in numerical order the r.h.s is non-negative and vanishes iff $P = P_s \,\forall\, r, s$ □

Proof of B. under these assumptions it is also true that

$$-\dot{S} = \sum_{r,s;r\neq s} U_{rs}(P_s - P_r) \ln P_r. \tag{142}$$

$$= -\frac{1}{2} \sum_{r,s;r\neq s} U_{rs}(P_{-}P_r)(\ln P_s - \ln P_r) \tag{143}$$

But

$$(x-y)(\ln x - \ln y) \geq 0. \tag{144}$$

□

The problem is that in general $U_{rs} \neq U_{sr}$. In fact

$$U_{rs} = |\langle r|T|s\rangle|^2, \tag{145}$$

where the S-matrix is given by

$$S = 1 + iT. \tag{146}$$

Unitarity them implies that

$$\sum_s U_{rs} = \sum_r U_{rs}. \tag{147}$$

9.11 Consequence of Symmetries

It is well known that in standard S-matrix quantum mechanics that that symmetries and conservation laws are closely related. In the case of \$-matrix quantum mechanics the connection is much less close.

9.12 S-Matrix Case

Wigner's theorem tells us that if T be a norm preserving map acting on the pure states preserving probabilities, then T must be unitary or anti unitary $T^{-1} = T^{\dagger}$. We also assume a similar map T' acts on the out pure states $T'^{-1} = T'^{\dagger}$, then if the S matrix is invariant

$$ST = T'S. \tag{148}$$

Thus

$$STS^{-1} = T'. \tag{149}$$

Now if

$$T = \exp i\epsilon G, \qquad G = g^{\dagger}, \tag{150}$$

then

$$SGS^{-1} = G', \tag{151}$$

where $T' = \exp i\epsilon G'$.

Thus if $|\text{out}\rangle = \$|\text{in}\rangle$

$$\langle \text{in}|G|\text{in}\rangle = \langle \text{out}|G'|\text{out}\rangle. \tag{152}$$

In other words, H is conserved. More over it also follows that any power G^k of G is conserved and that eigenstates of H and are taken to eigenstates of G'.

9.13 \$-Matrix Case

In the S-matrix case, a density matrix transforms under T as

$$\rho \to \mathcal{T}\rho = T\rho T^{\dagger}. \tag{153}$$

with

$$\mathcal{T} = \mathcal{T}^{\dagger}. \tag{154}$$

The condition of symmetry is now

$$\$ = \mathcal{T}'^{-1}\$\mathcal{T} = \mathcal{T}'^{\dagger}\$\mathcal{T}. \tag{155}$$

It is easy to see with particular examples that, in general symmetries, do not imply conservation laws [83].

10 Superscattering and the Reals

Hawking's original proposal (now famously abandoned by him) assumed the standard complex structure of quantum mechanics. From the point of view of what I have been advocating it seems curiously conservative to maintain that while advocating a much more radical modification of what we mean by the laws of physics.

In fact the entire discussion above works just as well over the reals, that is when the density matrices are just real symmetric semi-definite.

The general theory of super-scattering matrices works over all three fields, $\mathbb{R}$, $\mathbb{C}$ and $\mathbb{H}$ and interestingly the space of such matrices is itself a convex set. Now any convex set is, by a Theorem of Minkowski, the convex hull of its extreme points. In this case, the extreme points are unitary or anti-unitary purity preserving maps, i.e. S-matrices.

A simpler case to consider is restrict attention to the case of diagonal density matrices. In this case, \$ matrices are the doubly stochastic matrices encountered in the theory of Markov processes. These are the convex hull of the permutation matrices which take pure states to pure states.

The general theory of \$ matrices, at least in finite dimension, is nicely discussed in [84].

11 Conclusion

We have seen in this contribution that:

- Time and its arrow are intimately linked with the complex nature of quantum mechanics
- It is not difficult to construct spacetimes for which no arrow of time exists and on which backgrounds only real quantum mechanics is possible
- Only Riemannian manifolds admitting a reflection map Θ allow the recovery of standard quantum mechanics
- Even if one can define an arrow of time it may not be possible to define an operator Θ which reverses it

Why then do we have such a strong impression that time exists and that it has an arrow? When and how did the complex numbers get into quantum mechanics?

Like so many things in life: its all a matter of history. The universe "started" with very special initial conditions "when" neither time nor quantum mechanics were present. Both are emergent phenomena. Both are consequences of the special state we find ourselves in.

Constructing and understanding that state, and its alternatives is the on-going challenge of Quantum Cosmology.

Acknowledgements I would like to thank, Thibault Damour, Stanley Deser, Marc Henneaux and John Taylor for helpful discussions and suggestions about the material in Section 11.

Appendix: Complex Versus Real Vector Spaces

In this appendix we review some mathematical facts about complex structures. The standard structure of quantum mechanics requires that (pure) states are rays in a Hilbert space $\mathcal{H}_{\rm qm}$ which is a vector space over the complex numbers carrying a Hermitian positive definite inner product $h(U,V)$ such that

$$(i) \qquad h(U,\lambda V) = \lambda h(U,V), \qquad \forall \lambda \in \mathbb{C}. \tag{156}$$

$$(ii) \qquad h(U,V) = \overline{h}(V,U). \tag{157}$$

$$(iii) \qquad h(U,U) > 0. \tag{158}$$

It follows that $h(U,V)$ is antilinear in the first slot

$$h(\lambda U, V) = \overline{\lambda} h(U,V), \qquad \forall \lambda \in \mathbb{C}. \tag{159}$$

In Dirac's bra and ket notation elements of $\mathcal{H}_{\rm qm}$ are written as kets:

$$V \leftrightarrow |V\rangle \tag{160}$$

and elements of the $\mathbb{C}$- dual space $\mathcal{H}^\star_{\rm qm}$, the space of $\mathbb{C}$- linear maps $\mathcal{H}_{\rm qm} \rightarrow \mathbb{C}$ as bras: and there is an anti-linear map from $\mathcal{H}_{\rm qm}$ to $\mathcal{H}^\star_{\rm qm}$ given by

$$U \rightarrow \langle U| \tag{161}$$

such that

$$h(U,V) = \langle U \mid V \rangle, \tag{162}$$

thus

$$\langle U| = h(U,\cdot). \tag{163}$$

In components

$$|V\rangle = V^i \, |i\rangle \tag{164}$$

and

$$\langle U| = \langle j| \, \bar{U}^{\bar{j}}, \tag{165}$$

$$\langle U \mid V \rangle = h(U,V) = h_{\bar{i}j} \bar{U}^{\bar{i}} V^j, \tag{166}$$

where

$$h_{\bar{i}j} = \langle j \mid i \rangle, \tag{167}$$

and

$$\overline{h_{\bar{i}j}} = h_{\bar{j}i}. \tag{168}$$

Complex Vector Spaces As Real Vector Spaces

A useful references for this material with a view to applications in physics are [2,29].

For simplicity of exposition one may imagine that $\mathcal{H}_{\mathrm{qm}}$ as finite dimensional $\dim_{\mathbb{C}}\mathcal{H}_{\mathrm{qm}} = n < \infty$. Since a complex number is just a pair of real numbers [27], any Hermitian vector space may be regarded as a real vector space V of twice the dimension $\dim_{\mathbb{R}} V = 2n$ with something added [1], a complex structure J, i.e a real-linear map such that

$$J^2 = -1, \tag{169}$$

and a positive definite metric. g such that J is an isometry, i.e.

$$g(JX, JY) = g(X, Y). \tag{170}$$

It follows that V is also a symplectic vector space, with symplectic form

$$\omega(X, Y) = g(JX, Y) = -\omega(Y, X), \tag{171}$$

and J acts canonically, i.e.

$$\omega(JX, JY) = \omega(X, Y). \tag{172}$$

Alternatively given J and the symplectic form ω we obtain the metric g via

$$g(X, Y) = \omega(X, Jy). \tag{173}$$

The standard example is the complex plane $\mathbb{C} = \mathbb{R}^2$ where if

$$\mathbf{e}_1 = (1, 0), \qquad \mathbf{e}_2 = (0, 1), \tag{174}$$

$$J(\mathbf{e}_1) = \mathbf{e}_2, \qquad J(\mathbf{e}_2) = -\mathbf{e}_1 \tag{175}$$

or as a matrix

$$J = \begin{pmatrix} 0 & -1, \\ 1 & 0 \end{pmatrix} \tag{176}$$

and thus

$$J(x\mathbf{e}_1 + y\mathbf{e}_2) = x\mathbf{e}_2 - y\mathbf{e}_1 \tag{177}$$

which is the same in the usual notation as

$$i(x+iy) = -y + ix, \tag{178}$$

where $1 \leftrightarrow (1,0)$ and $i \leftrightarrow (0,1)$.

A complex structure J can be thought of as a rotation of ninety degrees in n orthogonal two planes. To specify it therefore it suffices to specify the (unordered) set of planes and the *sense* of rotation in each two-plane.

A Real Vector Space As a Complex Vector Space

Given the original real vector space, how are the complex numbers actually introduced? We start with V and pass to its *complexification*, the tensor product

$$V_{\mathbb{C}} = V \otimes_{\mathbb{R}} \mathbb{C}. \tag{179}$$

Note that $\dim_{\mathbb{R}} V_{\mathbb{C}} = 4n = 2\dim_{\mathbb{C}} V_{\mathbb{C}}$,

We now extend the action of J to $V_{\mathbb{C}}$, so it commutes with $i \in \mathbb{C}$:

$$J\alpha X = \alpha J X, \qquad \forall \quad \alpha \in \mathbb{C}, X \in \mathbb{C}. \tag{180}$$

We may now diagonalize J over $\mathbb{C}$ and write

$$V_{\mathbb{C}} = W \oplus \overline{W} \tag{181}$$

where

$$JW = iW, \qquad J\overline{W} = -i\overline{W}. \tag{182}$$

Clearly $\dim_{\mathbb{R}} W = 2n = 2\dim_{\mathbb{C}} W = \dim_{\mathbb{R}} V$, and W may be thought of as V in complex notation.

Thus if $X \in V$, we have that

$$X = \frac{1}{2}(1 - iJ)X + \frac{1}{2}(1 + iJ)X, \tag{183}$$

with $\frac{1}{2}(1 - iJ)X \in W$ and $\frac{1}{2}(1 - iJ)X \in \overline{W}$. Vectors in W are referred to as type $(1,0)$ or holomorphic and vectors in $\overline{W}$ as type $(0,1)$ or anti-holomorphic.

The Metric on $V_{\mathbb{C}}$

If V admits a metric for which J acts by isometries, we may extend the metric g to all of $V_{\mathbb{C}} = W \oplus \overline{W}$ by linearity over $\mathbb{C}$, we find that

$$(i)\ \ g(\bar{U},V)=\overline{g}(U,V) \tag{184}$$

$$(ii)\ \ g(U,\bar{U})>0, \tag{185}$$

$$(iii)\ g(U,V)=0, \forall U,V\in W, \text{and}, \forall U,V\in \overline{W}. \tag{186}$$

Negative Probabilities?

The metric g is usually assumed to positive definite because of the demand that probabilities be positive and lie in the interval [0, 1]. This requirement has been brought into question, notably by Feynman [28]. In the context of vacuum energy one should perhaps not be too quick in rejecting this possibility since the expectation value of the energy momentum tensor for negative probability states in such theories can of course have the opposite sign from the usual one. This could have applications the cosmological constant problem.

References

1. E.C.G. Stueckelberg, Quantum theory in real hilbert space. Helv. Phys. Acta. **33**, 727–752 (1960)
2. A. Trautman, On Complex Structure in Physics, in *On Einstein's path* (Springer, New York, 1996), pp. 487–501 [arXiv:math-ph/9809022]
3. F.J. Dyson, The threefold way: Algebraic structure of symmetry groups and ensembles in quantum mechanics. J. Math. Phys. **3**, 1199 (1962)
4. J.M. Jauch, *Foundations of Quantum Mechanics* (Addison Wesley, Boston, 1966)
5. J.M. Jauch, C. Piron, What is quantum logic. *Quanta* ed. by P.G.O. Freund (University of Chicago Press, Chicago, 1978) pp. 166–81
6. J.B. Barbour, Time and complex numbers in canonical quantum gravity, Phys. Rev. **D47**, 5422–5429 (1993)
7. J. Myheim, Quantum Mechanics on Real Hilbert Space [arXiv:quant-ph/9905037]
8. G.W. Gibbons The elliptic interpretation of black holes and quantum mechanics. Nucl. Phys. **B271**, 497–508 (1986)
9. G.W. Gibbons, H.J. Pohle, Complex numbers, quantum mechanics and the beginning of time. Nucl. Phys. **B 410**, 117 (1993) [arXiv:gr-qc/9302002]
10. A. Ashtekar, A. Magnon, Quantum fields in curved spacetime. Proc. Roy. Soc. London **A 346**, 375–94 (1975)
11. G.W. Gibbons, *The Einstein equations and the rigidity of quantum mechanics* lecture given at Quantum Mechanics and Cosmology a meeting in celebration of Jim Hartle's 60th birthday at the Isaac Newton Institute 2 Sept 1999 available on line at http://www.newton.ac.uk/webseminars/hartle60/3-gibbons/noframes.html, last accessed 30th April 2012
12. G.W. Gibbons, *Discrete Symmetries and Gravity* lecture given at the Andrew Chamblin Memorial Conference at Trinity College, Cambridge, Saturday 14th Oct 2006 available on line at http://www.damtp.cam.ac.uk/research/gr/workshops/Chamblin/2006/, last accessed 30th April 2012
13. J.L. Friedman, Lorentzian universes from nothing. Class. Quant. Grav. **15**, 2639 (1998)

14. J.R. Gott, X.I. Li, Can the universe create itself? Phys. Rev. **D 58**, 023501 (1998) [arXiv:astro-ph/9712344]
15. G.W. Gibbons, J.B. Hartle, Real tunneling geometries and the large scale topology of the universe. Phys. Rev. **D 42**, 2458 (1990)
16. G.W. Gibbons, in *Proceedings of the 10th Sorak School of Theoretical Physics*, ed. by J.E. Kim (World Scientific, Singapore, 1992)
17. A. Uhlmann, On Quantization in Curved spacetime. in Proceedings of the 1979 Serpukhov International Workshop on High Energy Physics; Czech. J. Phys. **B31** 1249 (1981); **B32** 573 (1982); Abstracts of Contributed Papers to GR9 (1980)
18. G.W. Gibbons, Quantization about Classical Background Metrics, in *Proceedings of the 9th G.R.G. Conference*, ed. by E. Schmutzer, (Deutscher, der Wissenschaften, 1981)
19. G.F. De Angelis, D. de Falco, G. Di Genova, Comm. Math. Phys. **103**, 297 (1985)
20. J. Glimm, A. Jaffe, A note on reflection positivity. Lett. Math. Phys. **3**, 377–378 (1979)
21. J. Frohlich, K. Osterwalder, E. Seiler, On virtual representations of symmetric spaces and their analytic continuation. Ann. Math. **118**, 461–481 (1983)
22. A. Jaffe, G. Ritter, Quantum Field Theory on Curved Backgrounds. II. Spacetime Symmetries, [arXiv:0704.0052 [hep-th]]
23. A. Jaffe, G. Ritter, Reflection positivity and monotonicity,' [arXiv:0705.0712 [math-ph]]
24. A. Jaffe, G. Ritter, Quantum field theory on curved backgrounds. I: The Euclidean functional integral. Commun. Math. Phys. **270**, 545 (2007) [arXiv:hep-th/0609003]
25. A. Chamblin, G.W. Gibbons, A judgment on sinors. Class. Quant. Grav. **12**, 2243–2248 (1995) [arXiv: gr-qc/9504048]
26. H.A. Chamblin, G.W. Gibbons, Topology and Time Reversal, in *Proceedings of the Erice School on String Gravity and Physics at the Planck Scale*, ed. by N. Sanchez, A. Zichichi [gr-qc/9510006]
27. W. Hamilton, Theory of conjugate functions, or algebraic couples; with a preliminary and elementary essay on algebra as the science of pure time. Trans. Roy. Ir. Acad. **17**, 293–422 (1837) Available online at http://www.maths.tcd.ie/pub/HistMath/People/Hamilton/, last accessd 30th April 2012
28. R. Feynman, Negative probability, in *Quantum Implications—Essays in Honour of David Bohm*, ed. by B.J. Hiley, F.D. Peat (Routledge & Kegan Paul, London, 1987) pp. 235–248
29. E.J. Flaherty, Hermitian and Kählerian Geometry in Relativity, Lecture Notes in Physics **46 (1976)**
30. G. 't Hooft, S. Nobbenhuis, Invariance under complex transformations and its relevance to the cosmological constant problem [gr-qc/0606076]
31. D.E. Kaplan, R. Sundrum, A symmetry for the Cosmological Constant [hep-hep/0505265]
32. S. Carlip, C. De-Witt-Morette, Where the sign of the metric makes a difference. Phys. Rev. Letts. **60**, 1599–1601 (1988)
33. J. Hucks, Global Structure of the standard model, anomalies, and charge quantization. Phys. Rev. **D 43**, 2709–2717 (1991)
34. S. Saunders, The negative energy sea, in *Philosophy of Vacuum*, ed. by S. Saunders, H. Brown (Clarendon Press, Oxford, 1991)
35. B. Grinstein, R. Rohm, Dirac and majorana spinors on non orientable Riemann surfaces. Commun. Math. Phys. **111**, 667 (1987)
36. A. Chamblin, G.W. Gibbons, A judgment on senors. Class. Quant. Grav. **12**, 2243–2248 (1995)
37. J.L. Friedman, Two component spinor fields on a class of time non-orientable spacetime. Class. Quant. Grav. **12** (1995)
38. A. Chamblin, On the obstruction to non-cliffordian pin structures. Comm. Math. Phys. **164**, 56–87 (1994)
39. R. Erdem, A symmetry for vanishing cosmological constant in an extra dimensional model. Phys. Lett. **B 621** (2005) [hep-th/0410063]
40. R. Erdem, C.S. Un, Reconsidering extra time-like dimensions [:hep-ph/0510207]

41. A.D. Linde, Appendix, Rep. Prog. Phys. **47**, 925 (1984)
42. A. Linde, Phys. Lett. **B 200**, 272 (1988)
43. A.D. Linde, Inflation, quantum cosmology and the anthropic principle, in *"Science and Ultimate Reality: From Quantum to Cosmos", honoring John Wheeler's 90th birthday*, ed. by J.D. Barrow, P.C.W. Davies, & C.L. Harper, Cambridge University Press (2003). e-Print: hep-th/0211048
44. G.W. Gibbons, A. Ishibashi, Topology and signature changes in braneworlds. Class. Quant. Grav. **21**, 2919 (2004) [hep-th/0402024]
45. J. Dorling, The dimensionality of time. Amer. J. Phys. **38**, 539 (1970)
46. G.W. Gibbons, D.A. Rasheed, Dyson pairs and zero-mass black holes. Nucl. Phys. **B 476**, 515 (1996) [hep-th/9604177]
47. M. Goldhaber, Speculations on Cosmogony. Science **124**, 218–219 (1956)
48. F.R. Stannard, The symmetry of the time axis. Nature **211**, 693–695 (1966)
49. M.G. Albrow, CPT conservation in the oscillating model of the universe. Nat. Phys. Sci. **241**, 3 (1973)
50. L.S. Schulman, Opposite thermodynamic arrows of time. Phys. Rev. Lett. **83**, 5419–5422 (1999)
51. J. Vickers, P.T. Landsberg, Thermodynamics: conflicting arrows of time. Nature **403**, 609–609 (2000)
52. L.-K. Hua, Geometry of symmetric matrices over any field with characteristic other than two. Ann. Math. **50**, 8 (1949); Causality and the Lorentz group. Proc. Roy. Soc. Lond. **A 380**, 487 (1982)
53. R.D. Schafer, *An Introduction to Nonassociative Algebras*, (Dover, 1995); **B253**, 573 (1985)
54. P.K. Townsend, The Jordan formulation of quantum mechanics: A review, in *Supersymmetry, Supergravity, and Related Topics*, ed. by F. del Aguila, J.A. de Azcárraga, L.E. Ibañez (World Scientific, Singapore, 1985)
55. G.W. Gibbons, Master equations and Majorana spinors. Class. Quantum Grav. **14**, A155 (1997)
56. A.D. Alexandrov, On Lorentz transformations. Uspekhi Mat. Nauk. **5**, 187 (1950)
57. E.C. Zeeman, Causality implies Lorentz invariance. J. Math. Phys. **5**, 490–493 (1964)
58. H.A. Chamblin, G.W. Gibbons, Topology and Time Reversal, in *Proceedings of the Erice School on String Gravity and Physics at the Planck Scale*, ed. by N. Sanchez, A. Zichichi [gr-qc/9510006]
59. G.W. Gibbons, The Kummer configuration and the geometry of Majorana spinors, in *Spinors, Twistors, Clifford Algebras and Quantum Deformations*, ed. by Z. Oziewicz et al. (Kluwer, Amsterdam, 1993)
60. G.W. Gibbons, How the Complex Numbers Got into Physics, talk at a one day seminar on *Complex Numbers in Quantum Mechanics*, Oxford 3 June (1995)
61. M. Dubois-Violette, Complex structures and the Elie Cartan approach to the theory of Spinors, in *Spinors, Twistors, Clifford Algebras and Quantum Deformations*, ed. by Z. Oziewicz et al. (Kluwer, Amsterdam, 1993)
62. E.L. Schucking, The Higgs mass in the substandard theory [arXiv:hep-th/0702177]
63. G.W. Gibbons, Changes of topology and changes of signature. Int. J. Mod. Phys. **D3**, 61 (1994)
64. M.J. Duff, J. Kalkkinen, Signature reversal invariance. Nucl. Phys. **B758**, 161 (2006) [arXiv:hep-th/0605273]
65. M.J. Duff, J. Kalkkinen, Metric and coupling reversal in string theory. Nucl. Phys. **B760**, 64 (2007) [arXiv:hep-th/0605274]
66. D. Singh, N. Mobed, G. Papini, The distinction between Dirac and Majorana neutrino wave packets due to gravity and its impact on neutrino oscillations [arXiv:gr-qc/0606134]
67. D. Singh, N. Mobed, G. Papini, Can gravity distinguish between Dirac and Majorana neutrinos? Phys. Rev. Lett. **97**, 041101 (2006) [arXiv:gr-qc/0605153]
68. J.F. Nieves, P.B. Pal, Comment on "Can gravity distinguish between Dirac and Majorana neutrinos?" Phys. Rev. Lett. **98**, 069001 (2007) [arXiv:gr-qc/0610098]

69. D. Singh, N. Mobed, G. Papini, Reply to comment on "Can gravity distinguish between Dirac and Majorana neutrinos?". Phys. Rev. Lett. **98**, 069002 (2007) [arXiv:gr-qc/0611016]
70. B. Zumino, Normal forms of complex matrices. J. Math. Phys. **3**, 1055–1057 (1962)
71. A. Peres, Gyro-gravitational ratio of Dirac particles. Nuovo Cim. **28**, 1091 (1963)
72. S.M. Cahn, *Fate, Logic and Time* (Yale University Press, New Haven 1967)
73. G.H. von Wright, *Time, Change and Contradiction* (Cambridge University Press, Cambridge, 1969)
74. G. Segre, There exist consistent temporal logics admitting changes of history [arXiv: gr-qc/0612021]
75. J.S. Thomsen, Logical relations among the principles of statistical mechanics and thermodymanics. Phys. Rev. **91**, 1263–1266 (195)
76. A. Aharony, Microscopic irreversibilty, unitarity and the H-theorem, in *Modern Developments in Thermodynamics*, ed. by B. Gal-Or (1974) pp. 95–114
77. A. Aharony, Y. Ne'eman, Time-reversal violation and the arrows of time. Lett. all Nuovo. Cimento. **4**, 862–866 (1970)
78. A. Aharony, Y. Ne'eman, Time-reversal symmetry violation and the oscillating universe. Int. J. Theor. Phys. **3**, 437–441 (1970)
79. Y. Ne'eman, Time reversal asymmetry at the fundamental level and its reflection on the problem of the arrow of time. in *Modern Developments in Thermodynamics* ed. by B. Gal-Or (1974) pp. 91–94
80. S.W. Hawking, Breakdown of predictability in gravitational collapse. Phys. Rev. **D 14**, 2460–2473 (1976)
81. R.M. Wald, Quantum gravity and time reversibility. Phys. Rev. **D21**, 2742–2755 (1980)
82. D.N. Page, Is black-hole evaporation predictable? Phys. Rev. Lett. **44**, 301–304 (1980)
83. D.J. Gross, Is quantum gravity unpredictable?, Nucl. Phys. **B 236**, 349 (1984)
84. M. Alberti, A. Uhlmann, *Stochasticity and Partial Order: Doubly Stochastic Maps and Unitary Mixing* (D Reidel, Dordrecht, 1982)
85. A.S. Eddington *The Mathematical Theory of Relativity* (Cambridge University Press, Cambridge, 1923) p. 25
86. G.W. Gibbons, A. Ishibashi, Topology and signature changes in braneworlds. Class. Quant. Grav. **21**, 2919 (2004) [arXiv:hep-th/0402024]
87. M. Mars, J.M. M. Senovilla, R. Vera, Lorentzian and signature changing branes. Phys. Rev. **D76**, 044029 (2007) [arXiv:0705.3380 [hep-th]]
88. M. Mars, J.M.M. Senovilla, R. Vera, Is the accelerated expansion evidence of a forthcoming change of signature?, Phys. Rev. **D77**, 027501 (2008) [arXiv:0710.0820 [gr-qc]]
89. L.-K. Hua, Causality and the Lorentz group. in *Proceedings of the Royal Society of London*; Ser A. Math. Phys. Sci. **380**, 487–488 (1982)
90. F.J. Yndurain, Disappearance of matter due to causality and probability violations in theories with extra timelike dimensions. Phys. Lett. **B 256**, 15–16 (1991)
91. I. Ya Arefeva, I.V. Volovich, Kaluza-Klein theories and the signature of spacetime. Phys. Lett. **164 B**, 287–292 (1985)
92. Y. Shtanov and V. Sahni, Bouncing braneworlds, *Phys. Lett.* **B 557**, 1 (2003) [arXiv:gr-qc/0208047]
93. J.G. Bennett, R.L. Brown, M.W. Thring, Unified field theory in a curvature-free five-dimensional manifold. Proc. Roy. Soc. **A 198**, 39–61 (1949)
94. I. Bars, C. Kounas, Theories with two times. Phys. Lett. **B 402**, 25 (1997) [arXiv:hep-th/9703060]
95. E.A.B. Cole, Prediction of dark matter using six-dimensional special relativity. Nuovo. Cim. **115 B**, 1149 (2000)
96. E.A.B. Cole, I.M. Starr, Detection of light from moving objects in six-dimensional special relativity, Nuovo. Cim. **105B**, 1091 (1990)
97. J.B. Boyling, E.A.B. Cole, Six-dimensional Dirac equation. Int. J. Theor. Phys. **32**, 801 (1993)
98. E.A.B. Cole, Six-dimensional relativity, in *Nagpur 1984, Proceedings, On Relativity Theory*, pp. 178–195

99. E.A.B. Cole, A proposed observational test of six-dimensional relativity. Phys. Lett. **A 95**, 282 (1983)
100. E.A.B. Cole, New electromagnetic fields in six-dimensional special relativity. Nuovo. Cim. **A 60**, 1 (1980)
101. E.A.B. Cole, Center-of-mass frames in six-dimensional special relativity. Lett. Nuovo. Cim. **28**, 171 (1980)
102. E.A.B. Cole, Particle decay in six-dimensional relativity. J. Phys. **A 13**, 109 (1980)
103. E.A.B. Cole, Emission and absorption of Tachyons in six-dimensional relativity. Phys. Lett. **A 75**, 29 (1979)
104. E.A.B. Cole, Subluminal and superluminal transformations in six-dimensional special relativity, Nuovo. Cim. **B 44**, 157 (1978)
105. E.A.B. Cole, Superluminal transformations using either complex space-time or real space-time symmetry. Nuovo Cim. **A 40**, 171 (1977)
106. B. Kostant, The Vanishing of Scalar Curvature and the Minimal Representation of $SO(4,4)$ in *Operator Algebras, Unitary Representations, Enveloping Algebras and Invariant Theory: Actes du colloque en l'honneur de Jacques Dixmier* Birkhauser, Basel (1990) 85–124
107. V. Guillemin, S. Sternberg, Variations on a theme of Kepler. AMS Collo. Publ. **42**, 72–73 (1990)
108. W.R. Biedrzycki, Einstein's equations with and embedding-dependent energy-momentum tensor for the compactified Minkowski time space and their relationship to the conformal action of $SO(4,4)$ on $S^3 \times S^3$. Proc. Natl. Acad. Sci. (US) **88**, 2176–2178 (1991)
109. W.R. Biedrzycki, Spinors over a cone, Dirac operator, and representations of $Spin(4,4)$. J. Funct. Anal. **113**, 36–64 (1993)
110. A.K. Das, E. Sezgin, Z. Khviengia, Selfduality in $(3+3)$-dimensions and the Kp equation. Phys. Lett. **B 289**, 347 (1992) [arXiv:hep-th/9206076]
111. S. Ferrara, Spinors, superalgebras and the signature of space-time [arXiv:hep-th/0101123]
112. L. Dabrowski, *Group Actions on Spinors* (Bibilopolis, Naples, 1988)
113. G.B. Halsted, Four-fold space and two-fold time. Science. **19**, 319 (1892)
114. J.B. Hartle, D.M. Witt, Gravitational theta states and the wave function of the universe, Phys. Rev. **D 37**, 2833 (1988)
115. N. Woodhouse, Geometric quantization and the bogolyubov transformation, Proc. Roy. Soc. Lond. **A 378**, 119 (1981)
116. A. Aguirre, S. Gratton, Inflation without a beginning: A null boundary proposal. Phys. Rev. **D 67**, 083515 (2003) [arXiv:gr-qc/0301042]
117. A.S. Wightman, On the localizability of quantum mechanical systems. Rev. Mod. Phys. **34**, 845–872 (1962)
118. W.G. Unruh, R.M. Wald, Time and the interpretation of canonical quantum gravity. Phys. Rev. **D 40**, 2598 (1989)
119. R. Giannitrapani, On a Time Observable in Quantum Mechanics [arXiv:quant-ph/9611015]
120. B.S. Kay, The Principle of locality and quantum field theory on (nonglobally hyperbolic) curved space-times." Rev. Math. Phys. **SI1**, 167 (1992)

The Phantom Bounce: A New Proposal for an Oscillating Cosmology

Katherine Freese, Matthew G. Brown, and William H. Kinney

Abstract An oscillating universe cycles through a series of expansions and contractions. We propose a model in which "phantom" energy with a supernegative pressure ($p < -\rho$) grows rapidly and dominates the late-time expanding phase. The universe's energy density is then so large that the effects of quantum gravity are important at both the beginning and the end of each expansion (or contraction). The bounce can be caused by high energy modifications to the Friedmann equation governing the expansion of the universe, which make the cosmology nonsingular. The classic black hole overproduction of oscillating universes is resolved due to their destruction by the phantom energy.

1 Introduction

The arrow of time is intimately connected to the entropy of the universe. The second law of thermodynamics inexorably drives us to ever increasing entropy, yet we live in neither a situation of maximal entropy (a black hole) nor in a minimal entropy universe. Apparently we thrive in the current "medium entropy" universe. How is this possible? In this volume, two possible explanations for this homogenous and isotropic universe we live in are especially discussed: special initial conditions or eternal inflation combined with anthropic arguments. In fact, there is a third option: cyclicity. Here the universe oscillates through a series of expansions and contractions. In a successful model of a cyclic nature, the entropy that has been

K. Freese (✉) · M.G. Brown
Department of Physics, Michigan Center for Theoretical Physics, University of Michigan, Ann Arbor, MI 48109, USA
e-mail: ktfreese@umich.edu; brownmattg@gmail.com

W.H. Kinney
Department of Physics, University at Buffalo, SUNY, Buffalo, NY 14260, USA
e-mail: whkinney@buffalo.edu

L. Mersini-Houghton and R. Vaas (eds.), *The Arrows of Time*, Fundamental Theories of Physics 172, DOI 10.1007/978-3-642-23259-6_7,

created in each cycle must yet again be destroyed in order to reset the stage for the next oscillation. We discuss a "phantom bounce" [1] as a proposal for an oscillating universe in which we postulate violation of the weak energy condition as a mechanism to destroy the (high entropy) black holes that are produced during each cycle. A key advantage of our proposal is that the phantom component of our proposal is testable in astrophysical data soon.

We discuss a scenario in which the universe oscillates through a series of expansions and contractions. After it finishes its current expanding phase, the universe reaches a state of maximum expansion which we will call "turnaround", and then begins to recollapse. Once it reaches its smallest extent at the "bounce", it will once again begin to expand. This scenario is distinguished from other proposed cyclic universe scenarios [2, 3] in that cosmological acceleration due to "phantom" energy (i.e., dark energy with a supernegative equation of state, $p < -\rho$) [4] plays a crucial role. In addition, our work differs from recent proposals in that our model takes place in three space and one time dimension (though the proposed mechanism for the bounce arises from braneworld scenarios). Since we originally proposed the phantom bounce in 2004, subsequent related work includes [5, 6]; see also [7]. Perturbations in this cosmology were discussed in [8].

The idea of an oscillating universe was first proposed in the 1930s by Tolman. Over the subsequent decades, two problems stymied the success of oscillating models. First, the formation of large scale structure and of black holes during the expanding phase leads to problems during the contracting phase [9]. The black holes, which cannot disappear due to Hawking area theorems, grow ever larger during subsequent cycles. Eventually, they occupy the entire horizon volume during the contracting phase so that calculations break down. (Only the smallest black holes can evaporate via Hawking radiation.) The second unsolved problem of oscillating models was the lack of a mechanism for the bounce and turnaround. The turnaround at the end of the expanding phase might be explained by invoking a closed universe, but the recent evidence for cosmological acceleration removes that possibility. For the observationally favored density of "dark energy", even a closed universe will expand forever. Thus, cyclic cosmologies appeared to conflict with observations.

Our scenario resolves these problems. Our resolution to the black hole overproduction problem is provided by a "phantom" component to the universe, which destroys all structures towards the end of the universe's expanding stage. Phantom energy, a proposed explanation for the acceleration of the universe, is characterized by a component Q with equation of state

$$w_Q = p_Q/\rho_Q < -1. \tag{1}$$

Since the sum of the pressure and energy density is negative, the dominant energy bound of general relativity is violated; yet recent work explores such models nevertheless. Phantom energy can dominate the universe today and drive the current acceleration. Then it becomes ever more dominant as the universe expands. With such an unusual equation of state, the Hawking area theorems fail, and black holes can disappear [10]. In "big rip" scenarios [11], the rapidly accelerating expansion due to this growing phantom component tears apart all bound objects including black holes. (We speculate about remnants of these black holes below.)

The phantom energy density becomes infinite in finite time [11, 12]. The energy density of any field described by equation of state w_Q depends on the scale factor a as

$$\rho_Q \sim a^{-3(1+w_Q)}. \tag{2}$$

Hence, for $w_Q < -1$, ρ_Q grows as the universe expands. Of course, we expect that an epoch of quantum gravity sets in before the energy density becomes infinite. We therefore arrive at the peculiar notion that quantum gravity governs the behavior of the universe both at the beginning and at the end of the expanding universe (i.e., at the smallest and largest values of the scale factor). Here we consider an example of the role that high energy density physics may play on both ends of the lifetime of an expanding universe: we consider the idea that large energy densities may cause the universe to bounce when it is small, and to turn around when it is large. The idea is economical in that it is the *same physics* which operates at both bounce and turnaround.

We use modifications to the Friedmann equations to provide a mechanism for the bounce and the turnaround that are responsible for the alternating expansion and contraction of the universe. In particular, we focus on "braneworld" scenarios in which our observable universe is a three-dimensional surface situated in extra dimensions. Several scenarios for implementing a bounce have been proposed in the literature [13, 14]. As an example, we focus on the modification to the Randall–Sundrum [16] scenario proposed by Shtanov and Sahni [13], which involves a negative brane tension and a timelike extra dimension leading to a modified Friedmann equation. Another example is the quantum bounce in loop quantum gravity [6, 15]. Once the energy density of the universe reaches a critical value, cosmological evolution changes direction: if it has been expanding, it turns around and begins to recontract. If it has been contracting, it bounces and begins to expand.

We emphasize that the two components we propose here work together: we use a modified Friedmann equation as a mechanism for a bounce and turnaround, and we add a phantom component to the universe to destroy black holes. Due to the phantom component, the same high energy behavior that produces a bounce at the end of the contracting phase also produces a turnaround at the end of the expanding phase. In addition, the bounce and turnaround are both nonsingular, unlike the cyclic scenario proposed by Steinhardt and Turok [3], which is complicated by a number of physical singularities related to brane collisions near the bounce [17]. This is currently a very controversial topic.

2 The Bouncing Cosmology

In an oscillating cosmology, what we observe to be "The Big Bang" really is the universe emerging from a bounce. The universe at this point has its smallest extent (smallest scale factor a) and largest energy density, somewhere near the Planck density. The universe then expands, its density decreases, and it goes through the classic radiation dominated and matter dominated phases, with the usual primordial nucleosynthesis, microwave background, and formation of large structure. A period of inflation may or may not be necessary to establish flatness and homogeneity. At a redshift $z = O(1)$, the universe starts to accelerate due to the existence of a vacuum component or quintessence field Q. We take a "phantom" component with $w_Q < -1$. The energy density of this component grows rapidly as the universe expands. Any structures produced during the expanding phase, including galaxies and black holes, are torn apart by the extremely rapid expansion provided by the phantom component. Any physics relevant at the high densities near the "Big Bang" again becomes important at the high densities near the end of the expanding phase. Modifications to the Friedmann equation become important at high densities, and cause the universe to turn around. The universe reaches a characteristic maximum density $2|\sigma|$ (which might be anywhere in the range from TeV to M_p), and starts to contract. As it contracts, at first its energy density decreases (as the phantom component decreases in importance), but then it again increases as matter and radiation become dominant. Eventually it reaches the high values at which the modifications to Friedmann equations become important. Once the energy density again reaches the same characteristic scale $2|\sigma|$, the universe stops contracting, bounces, and once again expands.

In the standard cosmology, there is no way to avoid a singularity for small radius or scale factor a. In the context of extra dimensions, however, one can have a bounce at finite a so that singularities are avoided. A nonsingular bounce is obtained if the Friedmann equation is modified by the addition of a new negative term on the right hand side:

$$H^2 = \frac{8\pi}{3M_p^2}\left[\rho - f(\rho)\right], \tag{3}$$

where the function $f(\rho)$ is positive. For a contracting universe to reverse and begin expanding again, we must have $\ddot{a} > 0$, which results in a condition on $f(\rho)$,

$$3(1+w)\rho f'(\rho) - 2f(\rho) - (1+3w)\rho > 0. \tag{4}$$

Similarly, for an expanding universe, $\ddot{a}$ must be negative for the expansion to reverse. A modified Friedmann equation of the form of (3) can be motivated in the context of braneworld scenarios, where our observable universe is a three-dimensional surface embedded in extra dimensions. Chung and Freese [18] showed that Einstein's equations in higher dimensions, together with Israel boundary conditions on our brane, can give rise to an equation of the form of (3). Different values of energy/momentum in the extra dimensions (the bulk) can be responsible for different $f(\rho)$ in (3).

In particular, we focus on "braneworld" motivated modifications to the Friedmann equation, where the modification to the Friedmann equation for the brane bound observer [13, 18, 19] is

$$H^2 = \frac{\Lambda_4}{3} + \left(\frac{8\pi}{3M_p^2}\right)\rho + \epsilon\left(\frac{4\pi}{3M_5^3}\right)^2\rho^2 + \frac{C}{a^4}, \tag{5}$$

where the last term (C is an integration constant) appears as a form of "dark radiation" (that is constrained like ordinary radiation), and ϵ corresponds to the metric signature of the extra dimension [13]. We will also assume that the bulk cosmological constant is set so that the three-dimensional cosmological constant Λ_4 is negligible.[1] Hence the relevant correction to the Friedmann equation is the quadratic term, $f(\rho) = \rho^2/2\,|\sigma|$. For $\epsilon < 0$, the Friedmann equation becomes

$$H^2 = \frac{8\pi}{3M_p^2}\left[\rho - \frac{\rho^2}{2|\sigma|}\right]. \tag{6}$$

One way to obtain $\epsilon < 0$ corresponds to an extra *timelike* dimension: models with more than one time coordinate typically suffer from pathologies such as closed timelike curves and non-unitarity. We use the model in [13] to motivate the choice of sign in the Friedmann equation, but a more detailed treatment would need to address these other issues to form a fully consistent picture.

Alternatively, in loop quantum gravity, there is a quantum bounce that takes place at Planck densities in lieu of the singularity in the standard classical Friedmann equation [6, 15]; if one couples this quantum bounce with a phantom component as in this paper, one would again obtain the same oscillating cosmology as discussed in this paper.

The expansion rate of the universe $H = 0$ at $\rho_{bounce} = 2|\sigma|$; it is at this scale that the universe bounces and turns around. For this choice of $f(\rho)$,

$$3(1+w)\rho f'(\rho) - 2f(\rho) - (1+3w)\rho = 3(1+w)\rho, \tag{7}$$

and the required condition on $\ddot{a}$ is satisfied at both bounce ($w > 0$) and turnaround ($w < -1$). On one end of the cycle it goes from contracting to expanding (this bounce looks to us like the Big Bang), and then at the other end of the cycle it goes from expanding to contracting. In models motivated by the Randall–Sundrum scenario, the most natural value of the brane tension is $\sigma = M_p$, but we treat the problem generally for any value of $\sigma >$ TeV.

At scales above $\rho > \sigma$, the validity of (6) breaks down in detail. However, the approach to $H = 0$ and thus the existence of a bounce and turnaround remain sensible. In any case, we use this braneworld model merely as an example of a correction to the Friedmann equation. Other modifications to the Friedmann

[1] This fine-tuning is the usual cosmological constant problem, which is not addressed in this paper.

equation might work as well, as long as there is the requisite minus sign in the equation.

3 Destruction of Black Holes

Black holes pose a serious problem in a standard oscillating universe. However, the Hawking area theorems that guarantee the continued existence of black holes have been constructed in special settings and may not apply here; e.g., the same modifications to gravity that give a bounce rather than a singularity in the cosmology may avoid singularities in the black holes. Indeed, when $w_Q < -1$, Davies [10] has shown that the theorem does not hold. Recently, Caldwell et al. [11] described the dissolution of bound structures in the "big rip" towards the end of a phantom dominated universe. Any black holes formed in an expanding phase of the universe are torn apart before they can create problems during contraction.

When are the black holes destroyed? We want to be certain that they are torn apart before turnaround. In general relativity, the source for a gravitational potential is the volume integral of $\rho + 3p$. An object of radius R and mass M is pulled apart when

$$-\frac{4\pi}{3}(\rho + 3p)R^3 \sim M. \tag{8}$$

Writing $\rho + 3p = \rho(1 + 3w_Q)$ during phantom domination and taking $R = 2GM$ for the black hole, we find that black holes are pulled apart when $-(4\pi/3)\rho(1 + 3w_Q)8M^3/M_p^6 \sim M$, which happens when the energy density of the universe has climbed to a value

$$\rho_{BH} \sim M_p^4 \left(\frac{M_p}{M}\right)^2 \frac{3}{32\pi} \frac{1}{|1 + 3w_Q|}. \tag{9}$$

More massive black holes are destroyed at lower values of ρ, i.e. earlier. It is the smallest black holes that get shredded last.

We must ensure that the black holes are destroyed before turnaround, so that $\rho_{BH} < \rho_{turn} = 2|\sigma|$. As an example, we can take $w_Q = -3$. Then 10^6 solar mass black holes, such as those at the centers of galaxies, get pulled apart when $\rho \sim 10^{-90} M_p^4$, which easily satisfies the above condition. The most tricky case would be Planck mass black holes, which either formed primordially or are relics of larger black holes that Hawking radiated. Even these should still be disrupted. From (9) these will be shredded when $\rho \sim 10^{-2} M_p$, before turnaround if the brane tension $|\sigma| = M_p^4$. However, for GUT scale brane tension $|\sigma| = m_{GUT}^4$, only black holes with $M \geq 10^5 M_p$ are disrupted. Fortunately these black holes Hawking evaporate in a time $\tau \sim (25\pi M^3/M_p^4)$ where M is the black hole mass. This occurs in only

$\sim 10^{-27}$ s for a black hole with $M = 10^5 M_p$. We also speculate that Planck mass remnant black holes that cannot disappear (still containing the singularity) may be dark matter candidates.

4 Discussion

Our proposal contains the novel feature that both bounce and turnaround are produced by the same modification to the Friedmann equation. However, it does so at the price of including more than one speculative element: the modified Friedmann equation requires a braneworld model to achieve, and the cosmology must be dominated by phantom energy. In many cases a phantom component is difficult to implement from a fundamental standpoint without severe pathologies such as an unstable vacuum (see, for example, [22].) However, Parker and Raval [23] have investigated a cosmological model with zero cosmological constant, but containing the vacuum energy of a simple quantized free scalar field of low mass, and found that it has $w < -1$ without any pathologies. Several additional areas also remain to be addressed. First, as the universe is contracting, those modes of the density fluctuations that we usually throw away as decaying (in an expanding universe) are instead growing. Hence dangerous structures may form during the contracting phase. At the end of the contracting phase, there is no phantom energy to wipe out whatever structure is formed. In this sense, the initial conditions for structure formation in this picture are set either during the phantom energy dominated epoch near turnaround or by the quantum generation of fluctuations in the collapsing phase [20]. Black hole formation could still kill the model. Second, it is not obvious that it is possible to create a truly cyclic (i.e. perfectly periodic) cosmology within the context of the "Phantom Bounce" scenario. The reason for this is entropy production. We speculate that it may be possible to create quasi-cyclic evolution by redshifting entropy out of the horizon during the period of accelerating expansion. Even more speculatively, we note that the special case of $w_Q = -7/3$, although disfavored by observation, possesses an intriguing duality between radiation ($\rho_{rad} \propto a^{-4}$) and phantom energy ($\rho_Q \propto a^4$). In this case, the behaviors of these components exchange identity under a transformation $a \to 1/a$ [21], effectively exchanging bounce for turnaround, a symmetry which might be exploited to achieve truly cyclic evolution.

On the observational side, our key ingredient is testable. The current expansion of the universe is the subject of much intense investigation. The universe is apparently accelerating, but the exact nature of the acceleration is not yet known. The previous value of the equation of state may be discovered over the next decade. The current uncertainty in the equation of state easily allows for the possibility of a phantom energy; some [24] have argued that $w_Q < -1$ is an excellent fit to the data. If upcoming observations discover that such a phantom energy indeed exists, then the community may be forced to conclude that the weak energy condition is violated and will need to rethink many basic assumptions. Phantom energy may be forced upon us, with the helpful consequence of permitting the "medium" entropy universe we inhabit.

References

1. M.G. Brown, K. Freese, W.H. Kinney, Bounce: A new oscillating cosmology, J. Cosmology Astropart. Phys. 0803, 002 (2008) [arXiv:astro-ph/0405353]
2. R. Tolman, *Relativity, Thermodynamics and Cosmology* (Oxford University Press, Oxford, 1934)
3. P. Steinhardt, N. Turok, Phys. Rev. **D65**, 126003 (2002); J. Khoury, P. Steinhardt, N. Turok, Phys. Rev. Lett. **92**, 031302 (2004)
4. S.M. Carroll, M. Hoffman, M. Trodden, Phys. Rev. D **68**, 023509 (2003); A. Melchiorri, L. Mersini, C.J. Odman, M. Trodden, Phys. Rev. D **68**, 043509 (2003)
5. L. Baum, P.H. Frampton (2006), [arXiv:astro-ph/0608138]; I. Aref'eva, P. H. Frampton, S. Matsuzaki, Proc. Steklov Inst. Math. 265, 59–62 (2009) [arXiv:0802.1294 [hep-th]]
6. J. Mielczarek, T. Stachowiak, M. Szydlowski, Phys. Rev. D **77**, 123506 (2008) [arXiv:0801.0502 [gr-qc]]
7. S. Nojiri, S.D. Odintsov, Phys. Lett. B **595**, 1 (2004) [arXiv:hep-th/0405078]
8. T.J. Battefeld, G. Geshnizjani, Phys. Rev. D **73**, 064013 (2006) [arXiv:hep-th/0503160]
9. R. Dicke, P.J.E. Peebles, in *General Relativity: An Einstein Centenary Survey*, ed. by S. Hawking and W. Israel (Cambridge University Press, Cambridge, 1979)
10. P. Davies, Ann. Poincare Phys. Theor. **49**, 297 (1988); E. Babichev, V. Dokuchaev, Y. Eroshenko [arXiv:gr-qc/0402089]
11. R. Caldwell, M. Kamionkowski, N. Weinberg, Phys. Rev. Lett. **91**, 071301 (2003)
12. R. Caldwell, Phys. Lett. **B545**, 23 (2002)
13. Y. Shtanov, V. Sahni, Phys. Lett. B **557**, 1 (2003)
14. P. Kanti, K. Tamvakis, Phys. Rev. D **68**, 024014 (2003); S. Foffa, Phys. Rev. D **68**, 043511 (2003)
15. A. Ashtekar, T. Pawlowski, P. Singh, Phys. Rev. D **74**, 084003 (2006) [arXiv:gr-qc/0607039]
16. L. Randall, R. Sundrum, Phys.Rev.Lett. **83**, 3370 (1999)
17. D. Lyth, Phys. Lett. **B526**, 173 (2002); J. Martin, P. Peter, N. Pinto Neto, D.J. Schwarz, Phys. Rev. D **65**, 123513 (2002); C. Gordon, N. Turok, Phys. Rev. D **67**, 123508 (2003); J. Martin, P. Peter, Phys. Rev. Lett. **92**, 061301 (2004); T.J. Battefeld, S.P. Patil, R. Brandenberger [arXiv:hep-th/0401010]
18. D. Chung, K. Freese, Phys.Rev. **D61**, 023511 (2000)
19. P. Binetruy, C. Deffayet, D. Langlois, Nucl.Phys. **B565**, 269 (2000); E.E. Flanagan, S. Tye, I. Wasserman, Phys. Rev. **D62**, 024011 (2000); C. Csaki, M. Graesser, C. Kolda, J. Terning, Phys. Lett. **B462**, 34 (1999); J. Cline, C. Grojean, G. Servant, Phys. Rev. Lett. **83**, 4245 (1999); L. Mersini, Mod. Phys. Lett. **A16**, 1583 (2001); W. Goldberger, M. Wise, Phys. Lett. **B475**, 275 (2000)
20. L.E. Allen, D. Wands Phys. Rev. D **70**, 063515 (2004) [arXiv:astro-ph/0404441]
21. M.P. Dabrowski, T. Stachowiak, M. Szydlowski, Phys. Rev. D **68**, 103519 (2003); L.P. Chimento, R. Lazkoz, Phys. Rev. Lett. **91**, 211301 (2003); J.E. Lidsey [arXiv:gr-qc/0405055]; L.P. Chimento, R. Lazkoz [arXiv:astro-ph/040551]
22. J.M. Cline, S. Jeon, G.D. Moore, Phys. Rev. D **70**, 043543 (2004) [arXiv:hep-ph/0311312]
23. L. Parker, A. Raval, Phys. Rev. Lett. **86**, 749 (2001)
24. A. Melchiorri, L. Mersini-Houghton, C.J. Odman, M. Trodden, Phys. Rev. D **68**, 043509 (2003) [arXiv:astro-ph/0211522]

Notes on Time's Enigma

Laura Mersini-Houghton

Abstract Scientists continue to wrestle with the enigma of time. Is time a dynamic or a fundamental property of spacetime? Why does it have an arrow pointing from past to future? Why are physical laws time-symmetric in a universe with broken time-reversal symmetry? These questions remain a mystery. The hope has been that an understanding of the selection of the initial state for our universe would solve such puzzles, especially that of time's arrow.

In this contribution, I discuss how the birth of the universe from the multiverse helps to unravel the nature of time and the reasons behind the time-reversal symmetry of our physical laws. I make the distinction between a local emerging arrow of time in the nucleating universe and the fundamental time with no arrow in the multiverse. The very event of nucleation of the universe from the multiverse breaks time-reversal symmetry, inducing a locally emergent arrow. But, the laws of physics imprinted on this bubble are not processed at birth. Time-reversal symmetry of laws in our universe is inherited from its birth in the multiverse, since these laws originate from the arrowless multiversal time.

1 Introduction

Time – the enigmatic building block of the cosmos – has stubbornly challenged natural philosophers and scientists over millenia. What is time? Why does it have an arrow? Why isn't time's arrow "DNA-ed" into our physical theories? Such

L. Mersini-Houghton (✉)
Department of Physics and Astronomy, UNC-Chapel Hill, NC, 27599-3255, USA

Department of Applied Mathematics and Theoretical Physics, Cambridge University, Cambridge, UK
e-mail: mersini@physics.unc.edu

L. Mersini-Houghton and R. Vaas (eds.), *The Arrows of Time*, Fundamental Theories of Physics 172, DOI 10.1007/978-3-642-23259-6_8,

basic questions that touch upon one of nature's most fundamental properties remain mysterious.

The complexity of time's mystery becomes more enticing within the multiverse framework. I have been advocating the necessity of viewing the cosmos as a multiverse since the advent of the landscape of string theory. The reason is: an investigation of why we started with this universe [1] necessarily leads to the question, "as compared to what other possible universes?" [2]. The investigation of the birth of our universe from the landscape multiverse studied in [1, 2, 6], and described briefly in the next section, shows that the selection of the initial states for universes born from the multiverse is governed by the dynamics of matter and gravitational degrees of freedom (D.o.F) and their entanglement with the background multiverse. Their birth is neither a special event nor is it occurring at a special moment. Nonequlibrium dynamics of these initial states leads to a superselection rule that picks only the high energy states as "survivor" universes. Since the progress with the puzzle of the selection of the initial state of the universe [1, 2, 6] and time's enigma are intertwined, then an extension of physics into the multiverse framework allows for deeper insights into a conceptual understanding of time.

In what follows, the fundamental time in the multiverse is distinct from the local time in the nucleating bubble universes. This letter argues that fundamental time does not have an arrow. But, that an arrow of time emerges only locally at the bubble location due to the breaking of time-reversal symmetry by the out-of-equilibrium correlations between various D.o.F's of the bubble entangled with the multiverse. Through this approach [1, 2] an arrow of time and physical laws with time reversal symmetry can be concomitant.

2 The Three Enigmas of Time and the Multiverse

Time's enigma is comprised of three basic questions: (a) Why do we have an arrow of time?; (b) What is time, fundamental or emergent?; (c) Why are physical laws time-symmetric, i.e. independent of the arrow of time?

The first question is closely related to the selection of the initial conditions of the universe. In Sect. 2.1 I argue that the arrow emerges at the moment of the bubble nucleation because the entanglement of the initial state with the multiverse and its nonequilibrium gravitational dynamics, create an information loss about the underlying reality. The information loss about the multiverse breaks the time-reversal symmetry at the bubble.

The second question is still open and debated. However, when the nature of time is treated within the multiverse framework, we may be in a position to draw more specific conclusions. Based on the conservation of the total information in the multiverse, the only two options left by the reversal-symmetry of this conservation are: either time is fundamental; or, it does not exist at all. I reason in Sect. 2.2 that time in the multiverse is fundamental rather than nonexistent.

Energy and information conservation lead to time-translation and time-reversal symmetries. That is, multiversal time becomes a fundamental building block of the cosmos. Symmetries ensure fundamental time has no direction, no beginning and no end. Fundamental time is not the same as the local time at the bubble nucleations since the latter is dynamic, it breaks reversal symmetry and experiences an emergent arrow.

We now have a way of addressing the third (and probably the toughest) question, the time-reversal symmetry of the physical laws in a universe where the reversal symmetry is badly broken. As discussed in Sect. 2.3, when treated in a multiverse framework, fundamental time is directionless and consequently physical laws inherit its time-reversal symmetry. Despite that reversal symmetry is broken for the local time by the bubble nucleation, the bubble still inherits laws of physics at birth from the multiverse, without modification. Thus the emergent time's arrow in the bubble does not affect the time-reversal symmetry imprinted onto the physical laws that the bubble inherits from birth in the multiverse.

This contribution offers a way of understanding the nature of time, the emergence of its arrow and the time-reversal symmetry of physical laws in a coherent picture, by posing time's enigma problem in the context of the multiverse.

2.1 Time's Arrow and the Birth of the Universe

We know what the universe looks like at present. We also experience an arrow of time from past to future. This arrow of time provides a profound insight into the initial moments of the universe. The reason is the second law of thermodynamics which leads us to conclude that time's arrow is a direct consequence of the asymmetry between the disorder of the present state and the order that must have existed in the initial state. More specifically, time's arrow implies that our universe had to start from a highly improbable state of exquisite order, and its equivalent low entropy. For this reason, the arrow is closely related to the mystery of the nucleation moment of the universe. In isolation, the second law of thermodynamics does not then resolve the enigmatic time's arrow problem but simply trades it with the enigma of what selected the initial state of the universe. But an understanding of the selection of the initial conditions of the universe would definitely represent progress in resolving the puzzle of the observed time's arrow in our universe. However, understanding time's arrow (a) is not sufficient since we still have to explore what time is (b), and why the physical laws are "unaware" of this arrow of time (c).

Exploring such questions requires a reconstruction of events from the present time to the Big Bang and before. As is well known, reverse-engineering is generically an ill-posed problem because a multiplicity of initial states can lead to a single present state. With this warning, even a sensible answer to time's enigma that relates it to the birth of the universe, carries a lot of ambiguity and remains in the realm of speculation until we can test the theory by experiment.

Nevertheless, exercising caution is useful for only as long as it does not discourage scientific inquiry. With this in mind, let us start investigating time's arrow by using the progress made in [1] for the selection of the initial conditions of our universe from the multiverse. A knowledge of the multiverse's structure would allow us to take a top-down approach and thus bypass reverse-engineeering ambiguity. In [1] we used the landscape derived from string theory as our working model for the multiverse structure. For the sake of illustration, let us continue our discussion of time using the same multiverse structure, the string theory landscape. The considerations below are applicable to other types, for example eternal inflation [13], if their structure is known and, crucially, if the selection rule for the surviving bubbles, (the measure), is governed by dynamics [14] instead of being fixed as an *apriori* initial condition.

The question: "why did our universe start in such a low entropy state" was investigated and addressed in [1] within the framework of the landscape multiverse. I will sketch briefly the main steps and results of this program since the selection of the initial conditions mechanism is directly relevant to the study of time's arrow here. The birth of the universe from the landscape multiverse in [1] was explored by proposing to place the wavefunction of the universe on the landscape multiverse, in order to study the dynamical evolution of matter and gravitational D.o.F's and their coupling to the multiverse "bath". The out-of-equilibrium dynamics in the initial states entangled with the multiverse "bath" leads to a superselection rule that eliminates the possibility of low energy initial states from the phase space, and selects only the highly ordered, high energy (low entropy) states as the most probable universes. The high energy states were dubbed "survivor universes" as they lead to the birth of physically relevant universes, and the low energy initial states were coined "terminal universes" as they can not give rise to expanding bubbles. The dynamics is contained in the Master Equation for the wavefunctional of the universe propagating on the multiverse. The Master Equation is a Schrödinger type equation with the gravitational and matter Hamiltonians being promoted to quantum operators. Thus it encaptures the dynamics of the wavefunctional of the universe and of the structure of the multiverse. But the Master Equation is sourced from a backreaction term of superhorizon matter modes acting on the wavefunctional. This term describes the entanglement of the multiverse "bath" with the wavefunction, which "pins down" the high energy branches of the wavefunctional, thereby triggering decoherence of our branch from the rest.

Locally this initial state is a "battlefield" that bubbles with the nonequilibrium dynamics of its matter and gravitational D.o.F's, along with the backreaction dynamics. The gravitational D.o.F.'s are captured by its vacuum energy which is trying to kick-start the initial bubble into an accelerated expansion. Entanglement with the multiverse and the backreaction of the matter D.o.F.s tries to crunch that initial state to a point. (In solid state jargon, the different behaviour of the two types of D.o.F's would be ascribed as follows: the matter D.o.F's constitute a "positive heat capacity" system while the gravitational D.o.F's constitute a "negative heat capacity" system. Thus the first type reaches equilbrium by driving to a crunch and the latter type reaches equlibrium by expanding to infinity. Having both types of dynamics drives the system out of equilibrium). Depending upon which one wins in this "tug-of-war" determines whether the initial packet survives and grows to give birth to a universe or terminates in a "stillbirth". The high energy states can survive the backreaction of matter and the bath, and can grow to physically relevant universes. But the low energy states can not survive. The superselection rule, derived from the nonequilibrium dynamics and entanglement with the multiverse "bath", selects the high energy states as the "survivor" universes and forbids the low energy "terminal" universes. The initial phase space of all possible states for potentially starting a universe like ours, thus shrinks to the subset of high energy initial states, the "survivor" universes. The main implication is that the phase space is not ergodic when dynamics is taken into account – an important point for the discussion of the dynamically driven asymmetry between the initial and boundary conditions below. The birth of the universe from the multiverse in this program thus offers the first explanation into the obstinate puzzle: why did our universe start in such a highly ordered (low entropy) state.[1] This resolution for the asymmetry between the entropy of the present universe, and the reasons behind the very low entropy of the initial state, then satisfactorily addresses the observed time's arrow puzzle.

Although this program [1,2,6] offers a natural explanation to one of the enigmas, the arrow of time, by facilitating our understanding of why the universe had to start in such an exquisitely ordered state of low entropy, within the multiverse framework, it is still an incomplete approach for the following reasons. The study of the dynamic evolution of the wavefunctional of the universe in the multiverse was carried out by implictly assuming the existence of time in the multiverse. That means that we still face two further questions in relation to understanding time, namely: (a) what is time in the multiverse? (b) why do our physical laws have a time-reversal symmetry instead of an arrow of time?

The question of what is fundamental time and the mystery of time's arrow are distinct, yet closely related. Completing the study for the arrow of time puzzle in [1] by using the Master Equation of quantum mechanics to study the dynamical evolution of the initial packet, now demands that we address the issue of the existence and the nature of multiversal time. Otherwise, until a tractable understanding of the nature of time is achieved, arguments presented here and in [1] would become circular.

2.2 *Fundamental Time and the Multiverse*

A useful way of thinking about entropy in cosmology is as a measure of the lack of information about the underlying reality. The underlying reality here is

[1]It is interesting that our program for the birth of the universe from the multiverse [1,6] has led to some intriguing observational consequences [6], with three of its predictions already succesfully tested so far, namely: the void [7], the dark flow [8], and σ_8 [4,5].

identified with the multiverse. Information is contained in physical correlations. Correlations are determined and quantified by physical laws. Then in principle, once the correlations are correctly identified, we should be able to estimate them.

In this discussion, energy and entropy are assumed to be meaningful concepts and, quantum mechanics is assumed to be valid should time exist. I will sometimes refer to the multiverse as the "bath", the nucleating universe as the "system" and time in the multiverse as "fundamental time". Let us now explore the question: "what is time in the multiverse", i.e. does it exist, does it make sense, does it have an arrow?

By definition the multiverse is all there is. Due to the unitarity principle, the total information of the "system" + "bath" is conserved. Since no information can be lost in the multiverse, then the only two consequent possibilites are: (a) *either time in the multiverse does not exist* [9]; (b) *or, time in the multiverse exists as a fundamental building block of the cosmos*, with no beginning, no end and with the reversability symmetry from the conservation of information.[2] Let us now explore where the first option, namely, time does not exist, leads: if time is nonexistent in the multiverse, then all the relevant physics to us is local rather then multiversal since time evolution and dynamics, would take meaning and emerge only at the bubble nucleation. With this choice, we have no need and no means of access to the underlying reality of the multiverse since a dynamic evolution would have no prior meaning or existence. This part of nature becomes redundant and irrelevant to a universe embedded in a *timeless* multiverse.

I will ascribe to the latter possibility, namely that fundamental time does exist in the multiverse because that part of reality is relevant and crucial for the birth of our universe. The necessity of the multiverse for understanding the birth of our universe, is based on the arguments presented in [1, 2], sketched in Sect. 2.1. Independently of time's enigma, the need for extending physics to the multiverse comes from basic questions such as: how did our universe come into being with such a special initial state. Such questions can not be meaningfully asked without the framework of the multiverse [1, 2]. Besides, the entanglement of our universe with its bath may have already proven its relevance by leaving testable imprints on astrophysical observations (see [7, 8] and footnote 1).

The existence of fundamental time in the multiverse becomes a logical conseqeunce when taking the view that the underlying reality, the bath in which our universe is a small domain, is relevant to our study of fundamental questions about nature. In fact, the opposite view that multiversal time may not exist, and the implication that the underlying reality of the multiverse is irrelevant, could lead to a "Loschmidt"-type paradox and obscure our understanding of entropy, time and arrow's emergence. Similar to the situation arising from the "molecular chaos" assumption, if multiversal time does not exist then local observers infer that the universe is a closed system with self-contained correlations. Such an assumption

---

[2]An emerging time in the multiverse does not appear plausible since the emergence adds information on the multiverse that wasn't there prior.

then leads to an information loss "sneaked in" by construction – by ignoring the information "hidden" in correlations between the multiverse and the universe, and in the gravitational sector – thereby creating an artificial, instead of a physical, asymmetry between initial and boundary conditions [11].

The view that the *multiverse is a closed system* but the *universe is an open system* entangled with the multiverse bath naturally leads to the second option, namely: fundamental time exists. Then, conservation of information in the multiverse results in the reversability symmetry of fundamental time. Which implies, time in the multiverse is arrowless. Universally laws of physics carry this time-reversal symmetry. Energy conservation would imply time-translation symmetry. This option leads us to conclude that multiversal time is fundamental, it has no direction, no beginning and no end.

Local time at the position of the nucleating universe, although related, is not the same as the fundamental time of its underlying bath. Entanglement with the multiverse and the coupling between the matter and gravitational D.o.F's, mentioned in Sect. 2.1 and derived in [1, 6], drives a dynamical evolution of correlations. That is, the universe is an open and out-of-equlibrium system. Initially, the wavepacket has a superposition of geometries. As the bath "pins down" the branches it entangles with, (the system decohering), then there is a flow of information not only between the matter and gravitational sectors but also to the multiverse. This information is contained in the off-diagonal terms of the reduced density matrix for our branch of the wavefunction that describes how fast the superposition of different gemoetries decohere from each-other, as a result of entanglement with the bath [1]. Other channels of information loss are given by the intrinsic interaction of matter with gravitational D.o.F.s, such as, particle creation from curved spacetime, which describes a transfer of information from the varying gravitational fields with zero entropy to the matter sector, as well as the generic coupling of matter to curvature, (gravity), contained in Einstein equations. These channels contain the excitations of the gravitational vacuum correlated nonlinearly to matter. Despite some intriguing attempts [15], the issue of gravitational entropy and its information transfer to the particle sector is still elusive and will be discussed in a subsequent paper. From the local observers point of view, more and more correlations "hide" as irrelevant when the bubble goes through the nonequlibrium dynamics of expansion and decoherence. The information is lost to the bath and the gravitational sector. As the universe grows, local observers in the branch continue to lose information about the underlying reality, which breaks the reversal symmetry of time locally – the "hidden" information is contained in the entanglement, information about the fact that this bubble is part of a bigger phase space, the multiverse. From the bubble's perpective, the information is lost in a non-reversible way due to the local nonequlibrium dynamics and decoherence for reasons described next. Such dynamics guarantees that the untangling of our branch from the multiverse bath does not occur, thus the irreversability of the process.

Time's arrow emerges only locally because time reversal symmetry is broken locally only, at the bubble nucleation, although the fundamental time of the multiverse is arrowless. If local observers were able to move away from the trajectory of the entangled and decohered branch, they would find time-symmetry restored away from the bubble. Thus, local breaking of time-reversal symmetry is due to the correlation changes between the system and the bath, which is driven by the gravitational dynamics of the system. Such local nonequilibrium, irreversable dynamics induces an asymmetry between the starting point, the initial state of the universe on the multiverse and the final state of the system. The system undergoes nonequlibrium dynamical evolution, which renders *its phase space nonergodic*, and ensures that the system can *never return to its initial state* [1]. We perceive this change in the correlations of the system from the bath as a separation of the system from the multiverse and deduce locality since the observer in the system defines the relevant degrees of freedom locally.

In the birth of the universe from the multiverse scenario [1, 2, 6], the initial conditions are not "hand-picked". Rather, they are dynamically superselected from a generic set. The high energy, out-of equlibrium, initial system then tries to drive towards a symmetric final state, thereby driving an increase in entropy. The superselection rule for the initial states derived in [1], separates "terminal" from "survivor" universes, by wiping out the former from the phase space available to our initial states. As a consequence of the superselection arising from the nonequilibrium dynamics and entanglement to the multiverse, phase space is not ergodic and Poincaré recurrences do not occur. The implication is that the system can never return to its initial state. Thus the symmetry between the initial and final state can not be restored, resulting in an emerging arrow in the bubble. The asymmetry between the initial and boundary conditions, in this theory [1], is not an artifact of breaking the symmetry by placing arbitrary conditions on the initial state while ignoring the boundary conditions, as rightly critized by H. Price [11]. The asymmetry is governed and driven by the superselection rule on the multiverse arising from nonequlibrium dynamics of matter and gravitational D.o.F.'s. This reasoning remains valid for contracting universes thus a reversal of time's arrow during the transition from an expanding to a contracting phase in an open system, such as the universe, also can not occur.

2.3 *Time-Reversal Symmetry for the Laws of Physics*

Although fundamental time in the multiverse from which the universe nucleated, has no arrow, the local observer experiences that an arrow of time has emerged at the bubble due to the information of the entanglement with the bath lost and hidden to the gravitational sector. At the bubble, time reversal symmetry is broken by the very act of nucleation and entanglement with the bath, since there is information loss about the underlying reality of the multiverse and about the gravitational entropy. As the bubble decoheres such entanglement with the bath is deemed as irrelevant, and these correlations ignored. The initial state is selected dynamically by the underlying physical laws. The nonequlibrium dynamics also ensures the nonergodicity of phase space which induces an asymmetry between the initial and boundary conditions – the bubble can never return to its initial state.

However, the system inherits the same laws of physics from the multiverse that were valid before its decohering, without processing or changing them. But since globally fundamental time has no arrow, and the laws of physics therefore are symmetric with respect to time reversal operations, then each nucleating universe inheriting these laws from birth to a multiverse, would carry the same time-reversal symmetry for their laws. The time reversal symmetry of laws is a direct consequence of the fundamental time in the multiverse and not the local time in the bubble. For this reason, the emerging arrow of time at the bubble location and the time-symmetry of the laws of physics are concomitant since they are independent in origin. Unlike the arrow of time, laws do not emerge at the bubble nucleation, they are inherited from the underlying theory. This addresses our second mystery in a cohesive way: the physical laws a universe is born with, can be time-symmetric despite the breaking of this symmetry locally that induces the emergence of the local time's arrow. The time-symmetry of laws inherited by "survivor" universes does not imply that physical laws are the same in every bubble. It simply makes a statement about a feature they all have in common from their origin in the multiverse, they share the time-symmetry property.

Here we have demonstrated how, within a multiverse framework, we can achieve a coherent existence of both phenomena, an emerging local arrow of time and time symmetric laws. The separation of the system from the bath produces an arrow of time but does not modify or process the physical laws.

3 Discussion

The situation with time's enigma and the reversal symmetry of physical laws is similar to the resolution of the Loschmidtz paradox concerning Boltzmann's H-theorem. The reason for the entropy increases in Boltzmann's approach was the assumption of "molecular chaos", which ignored the correlations and information about interaction and the microdynamics of particles. A similar situation arises in our case. Based on classical results of [16] that assume equlibrium for the closed system, the gravitational entropy is usually taken to be zero, except for objects with horizons, such as DeSitter geometries and black holes. Yet the entropy of particles created from these gravitational fields in the universe is not zero. Besides, the interaction between the matter and gravitational sector is always present, which ensures that the open system remains out-of-equlibrium. The transfer of information from the particle sector to the gravitational sector, for the open system immersed in a bath, results in a loss since the role and the nonequilibrium dynamics of gravitational degrees of freedom is not taken into account. The assumption of independence of the system from the multiverse bath, together with information transferred to the gravitational sector and contained in the gravitational entropy are ignored as irrelevant. Information lost via these channels by enforcing locality, equilibrium, and choosing local matter D.o.F.'s as the only relevant D.o.F's, breaks time-reveral symmetry and leads to the emergence of a local arrow of time. In [1, 2, 6] we showed how to incorporate the superhorizon nonlocal entanglement with the bath, into the system's correlation. A general approach to quantifying correlations and information loss to the gravitational sector would require an understanding of the role of gravitational D.o.F's to Boltzmann's kinetic equation and H-function. Once we have a handle on the information lost via correlations of the particle with the gravitational sectors, we would be in a position to test the theory. The coupling between the matter and gravitational sectors results in nonequilibrium dynamics with the rate of information transferred from one sector to the other providing a concept of clocks. Clocks could not be build in a universe in perfect equlibrium, such as the thermal bath of radiation of a pure DeSitter geometry since in this case thermal equilibrium requires that entropy remains a constant, and that particle creation from the gravitational vacuum excitations becomes extinct.

A universe nucleating from the multiverse through the dynamic selection of its initial conditions results in the following picture: time in the multiverse is arrowless, with no beginning, no end and it is a fundamental building block. Laws of physics inherit the time-symmetry of the underlying theory. But local time in the bubble universe is a dynamic parameter with an emerging arrow at birth, since more and more information is lost about the underlying reality and transferred to the gravitational sector, which creates an asymmetry between the initial and final states of the local universe. The bubble inherits the arrowless physical laws despite it breaking the time-symmetry. The initial conditions of the bubble are dynamically chosen from physical laws from a generic state. The bubble tries to drive towards symmetric boundary conditions. Therefore the emerging arrow is not a consequence of an artificially imposed symmetry breaking between the selection of a preferred initial condition, without the selection of a boundary conditions [11]. Initial and boundary conditions are both governed and dynamically selected by physical laws from generic sets, which can be achieved when the birth of the universe is studied within a multiverse framework with a fundamental arrowless time. The asymmetry arises from the information loss to the universe and the nonequlibrium dynamics of matter and gravitational D.o.F's that leads to a nonergodic phase space. Such asymmetry renders local time to be dynamic and have an emerging arrow (Fig. 1).

As I tried to caution at the start, any attempts at tackling time's enigma remain in the realm of speculation until the calculational tools of information transfer and gravitational entropy are discovered. Without these tools it is hard to make testable predictions of the theory since the information contained in the interactions between matter and gravitational vacuum can not be estimated. Yet, the theory described in this contribution, for time's enigma in the context of the dynamically selected birth of the universe from the multiverse, provides a coherent picture of the concomitant co-existence of the three aspects of this enigma: a locally emerging time's arrow, from a fundamentally arrowless time, for a universe that inherits the time-symmetry of its laws from the multiverse.

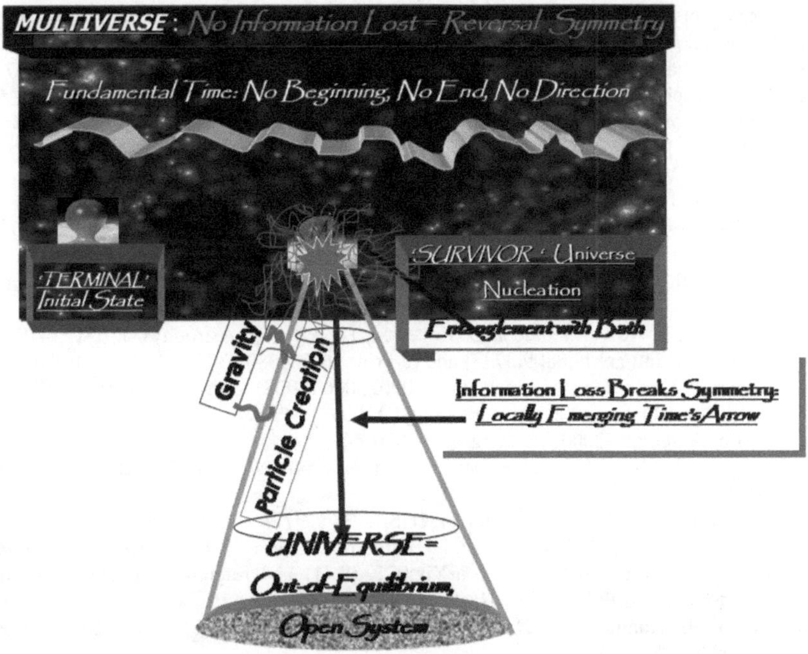

Fig. 1 A schematic drawing of the birth of the universe from the multiverse. Only the high energy initial states in the multiverses are dynamically selected to give borth to a universe. Low energy states become "terminal". Information is conserved in the multiverse thus multiversal time is fundamental and directionless. The nucleating universe is an open system. Its out-of-equlibrium dynamics and entanglement with the multiverse break the time-reversal symmetry locally. But physical laws of the bubble originate from the multiverse thus carry the reversal-symmetry.

Acknowledgements L.M-H is grateful to DAMTP for their hospitality during the time this work was done. LM-H is supported in part by DOE grant DE-FG02-06ER1418, NSF grant PHY-0553312 and FQXI.

References

1. R. Holman, L. Mersini-Houghton, Phys. Rev. **D74**, 123510 (2005) [e-Print: hep-th/0511102] and [e-Print: hep-th/0512070]; L. Mersini-Houghton, Class. Quant. Grav. **22**, 3481–3490 (2005) [e-Print: hep-th/0504026]; A. Kobakhidze, L. Mersini-Houghton, Eur. Phys. J. **C49**, 869–873 (2004) [e-Print: hep-th/0410213]
2. L. Mersini-Houghton, Sep 2008 [e-Print: arXiv:0809.362]; L. Mersini-Houghton, AIP Conf. Proc. **878**, 315–322, Madrid (2006), The dark side of the universe, 315–322, [e-Print: hep-ph/0609157]; L. Mersini-Houghton, AIP Conf. Proc. **861** 973–980, Paris (2005), Albert Einstein's century, 973–980, [e-Print: hep-th/0512304]
3. L. Mersini-Houghton, Selection of Initial Conditions: The Origin of Our Universe from the Multiverse, in: R. Vaas (ed.): *Beyond the Big Bang*. Springer, Heidelberg 2012 [e-Print: arXiv:0804.4280]

4. E. Komatsu et al. "Five-years Wilkinson microwave anisotropy probe (WMAP) observations: Cosmological interpretation" Astrophys. J. Suppl. 180, 330 (2009) [astro-ph/0803.0547]
5. By SDSS Collaboration, J.K. Adelman-McCarthy et al., Astrophys. J. Suppl. **175**, 297–313 (2008) [astro-ph/0707.3413]; M.E.C. Swanson, M. Tegmark, M. Blanton, I. Zehavi, Mon. Not. Roy. Astron. Soc. **385**, 1635–1655 (2008) [astro-ph/0702584]; R.R. Gibson, W.N. Brandt, D.P. Schneider [astro-ph/0808.2603]
6. R. Holman, L. Mersini-Houghton, T. Takahashi, Phys. Rev. **D77**, 063510 (Nov 2006) [e-Print: hep-th/0611223]; R. Holman , L. Mersini-Houghton, T. Takahashi, Phys.Rev. **D77**, 063511, (Dec 2006) [e-Print: hep-th/0612142]; L. Mersini-Houghton, R. Holman, J. Cosmology Astropart. Phys. **0902** 006 (2008) [e-Print: arXiv:0810.5388]
7. L. Rudnick, S. Brown, L.R. Williams, Astrophys. J. **671**, 40–44 (2007) [e-Print: arXiv: 0704.0908]
8. A. Kashlinsky, F. Atrio-Barandela, D. Kocevski, H. Ebeling, Astrophys. J. **691**, 1479–1493 (Sep 2008) [e-Print: arXiv:0809.3733] and [e-Print: arXiv:0809.3734]; L. Mersini-Houghton,, R. Holman, J. Cosmology Astropart. Phys. **0902**, 006 (2008) [e-Print: arXiv:0810.5388]
9. C. Kiefer, Does time exist in quantum gravity? (2009) [e-Print: arXiv:0909.3767]
10. L. Mersini-Houghton (2006) [arXiv:gr-qc/0609006]
11. H. Price, *Time's Arrow and Archimedes'Point: A View from Nowhen* (Oxford University Press, Oxford, 1996)
12. G.W. Gibbons, S.W. Hawking, Phys. Rev. **D15**, 2738 (1977)
13. A. Aguirre Eternal Inflation: Past and Future, in: R. Vaas (ed.): *Beyond the Big Bang*. Springer, Heidelberg 2012 [e-Print: arXiv:0712.0571] and references herein; S. Winitzki, Eternal inflation. World Scientific, Hackensack 2009
14. D. Simon, J. Adamek, A. Rakic, J.C. Niemeyer, JCAP 0911, 008 (2009) [e-Print: arXiv:0908.2757]
15. L. Smolin, Gen. Rel. Grav. **17**, 5 (1985); R. Brandenberger, T. Prokopec, V. Mukhanov (1992) [e-print: gr-qc/9208009]
16. G.W. Gibbons, S.W. Hawking, Phys. Rev. D., **15**, N.10:2752–2756; P.C.W. Davies, L.H. Ford, D.N. Page, Phys. Rev. D. **34**, N.6:1700–1706

A Momentous Arrow of Time

Martin Bojowald

Abstract Quantum cosmology offers a unique stage to address questions of time related to its underlying (and perhaps truly quantum dynamical) meaning as well as its origin. Some of these issues can be analyzed with a general scheme of quantum cosmology, others are best seen in loop quantum cosmology. The latter's status is still incomplete, and so no full scenario has yet emerged. Nevertheless, using properties that have a potential of pervading more complicated and realistic models, a vague picture shall be sketched here. It suggests the possibility of deriving a beginning within a beginningless theory, by applying cosmic forgetfulness to an early history of the universe.

1 Introduction

Time in quantum theory is something of a black sheep. In non-relativistic quantum mechanics it remains a classical parameter labelling the evolution of states, but is not allowed to fluctuate as position does. Even in relativistic systems and quantum field theory, time often appears as a disciplined parameter trained to order events, much as it is used in classical physics. Crucial for particle physics is the direction time provides for interaction events scattering initial states into final ones. But any directedness is simply put into the formalism. At the level of elementary reactions, time knows no order: if a reversal of time were allowed, events would still occur in

M. Bojowald (✉)
Institute for Gravitation and the Cosmos, The Pennsylvania State University, 104 Davey Lab, University Park, PA 16802, USA
e-mail: bojowald@gravity.psu.edu

L. Mersini-Houghton and R. Vaas (eds.), *The Arrows of Time*, Fundamental Theories of Physics 172, DOI 10.1007/978-3-642-23259-6_9,

any way, nearly unchanged.[1] It is only our choice of initial and final states which determines a scattering amplitude.

All this is different on the macroscopic level and especially in cosmology. Here, structures change with a trend. One often thinks of a simple initial state evolving into complexity, a puzzle to be explained by an arrow of time. If this is to be derived rather than postulated, a theory of initial states is required.

The main part of this contribution will be an exploration of the possibility that true quantum degrees of freedom, those such as fluctuations which completely lack a classical analog, could play a role of or for time. In this way, we will take seriously quantum space-time, not (just) as a new and possibly discrete structure but as a fully dynamical quantum entity. More specifically, cosmological models will lead us to an analysis of quantum correlations as quantities changing with a trend. If consistently realized, such a perspective is very different from the traditional ones regarding time: time would be inherently quantum; it would not exist in a classical world. In semiclassical physics, it remains only as a shadow of the quantum physics that lies beneath.

We will take advantage of a useful description of quantum dynamics (sketched in the Appendix) based on the evolution of characteristic quantum variables, rather than whole but partially redundant wave functions. The same kind of description can be used to explore the nature of non-singular big bangs. Such events, while still playing the role of the moment of commencement for the part of the universe accessible to us, can no longer be viewed as entire initial states of the universe: with the singularity being resolved, there is a universe before the big bang. But specific realizations of such scenarios do have derived features of special initial states as they may be posed at the big bang. In this way, dynamical properties give insights into the question of initial states and the directed evolution that ensues. Especially the phenomenon of cosmic forgetfulness shows that much of the state before the big bang remains hidden after the big bang. Without remembrance, the arrow of time might well be considered blunt – or do we just see the blunt end of a reversed arrow?

By its nature, our analysis will be incomplete and preliminary. No clear scenario emerges yet; just several indications exist. But they may show that the topics touched here are still worth pursuing.

2 The Problem and the Arrow of Time

Many questions are to be addressed in the context of time. The most important one is, of course, the aptly named problem of time [2]. It arises mainly in canonical formulations of gravity and attempted quantizations, but its nature reaches farther. Independently of technical aspects, it is about the question whether there is an unambiguous degree of freedom in generally relativistic theories which can play

[1] The laws are, of course, not completely time reflection symmetric, which might be exploited in the context discussed here [1].

the role of time, or of a parameter whose values arrange causally related events and thus separate the past, present and future.

This is to be distinguished from the question of the arrow of time [3], which irreversibly orders events already separated into past, present and future by the time variable. Such an arrow is often related to thermodynamical questions via entropy, or to the selection of special initial states in quantum cosmology. The question of the arrow of time builds on an existing time variable and is thus to be separated from the problem of time, which is more basic. In this contribution, we start with a discussion of the problem of time.

The problem initially arose in canonical quantizations of gravity, with a dynamics governed only by a Hamiltonian constraint, not by a true Hamiltonian. Thus, quantum states do not seem to evolve, quite obviously in conflict with the perception of change.[2] Without any notion of space-time coordinates in canonical quantizations of gravity, which provide operators for geometrical quantities derived from the space-time metric but nothing for coordinates, the usual way out by coordinatizing time is blocked. One is forced to identify an appropriate time degree of freedom from the physical variables, such as geometrical ones or matter fields. The problem is that none of them seems to be a globally valid choice for time as an unambiguous labelling of events.

While this problem becomes technical and pressing in canonical quantum gravity, it is more general as well as deeper than might be indicated at first sight. If we were able to identify a global time variable from the physical degrees of freedom, we would be led to attributing a new physical meaning to time. Time would cease to be a conventional description of observed change and become a physical quantity on par with all others. It would be subject to physical laws, and would fluctuate in quantum theories. In that case, one might as well look for a global time variable among the true quantum degrees of freedom of a relativistic system, a degree of freedom such as quantum fluctuations or correlations without a classical analog. From a dynamical systems perspective, these are degrees of freedom in their own right. (Such variables do play a special role from the perspective of quantum observables since they are not obtained through expectation values of one linear operator. But quantum fluctuations, for instance, are certainly measurable in the same statistical sense as expectation values.) The fundamental notion of time would crucially be tied to quantum physics, while classical physics would have to resort to time coordinates, a poor substitute for a truly deep notion.

---

[2]It is interesting to note that the problem of time and motion becomes pressing when quantum gravity is considered. Quantum gravity is often tied to another expectation, that of discreteness or an atomic nature of space-time. Maybe solving the problem of time would lead us to establishing an atomic nature of time? If so, this would be reminiscent of a much older debate among pre-sokratic philosophers: Parmenides denied any reality to motion and change, which he logically argued to be pure illusion. His most basic statement was that nothingness does not exist, and so a body cannot move from where it is now to a place of empty space which was thought not to exist. The logical conflict was resolved by the atomists who accepted the notion of empty space and were led to the concept of material atoms.

Several indications exist for the squeezing of quantum matter [4–7] or gravitational waves [8] to play the role of time. Here, following suggestions in [9, 10] we describe results to explore a possible relation to quantum gravity states,[3] indicating an emergent concept of time in a quantum description of universe models. If these models and ideas are correct, the quantity ultimately playing the role of time is not the one put in initially to set up the evolution equations, and it is not the one used in a classical description. This quantity, the true nature of time in the picture proposed, does not at all exist in the classical theory. So far, these considerations are inconclusive concerning the problem of time. But the methods will set the stage for a discussion of the arrow of time.

3 Classical Dynamics

We start with a cosmological system where the problem of time is solved trivially: a model sourced by a free, massless scalar field ϕ. Thanks to the absence of any non-trivial potential, the value of the scalar is monotonic in any time coordinate and thus can itself be used as time. While these models are rather simple, some exactly solvable versions provide a basis for a much more general analysis.

With a free, massless scalar as the sole matter content of an isotropic universe with cosmological constant Λ, the expansion history, for the different choices of spatial curvature via $k = 0$ or $k = \pm 1$, is determined by the Friedmann equation

$$\left(\frac{\dot{a}}{a}\right)^2 + \frac{k}{a^2} = \frac{4\pi G}{3} \frac{p_\phi^2}{a^6} + \Lambda$$

for the scale factor a. The dot denotes a derivative by proper time, leading to the Hubble parameter $\dot{a}/a$. The coupling to matter is quantified by the gravitational constant G, multiplying the energy density of matter. Here, for a free, massless scalar, only kinetic energy is contributed via the momentum $p_\phi = a^3\dot{\phi}$. In what follows, we will use $k = 0$ and $\Lambda < 0$ to be specific, though not realistic. (The case of a positive cosmological constant is very similar to the negative sign as far as classical dynamics is concerned, but is much more subtle at the quantum level. One can find hints of this subtlety in the existence of different self-adjoint extensions of the quantum Hamiltonian [12] or in the dynamical behavior of quantum states.)

To employ canonical quantization later on, we now introduce the classical canonical formulation. Choosing the (rescaled) volume $V = a^3/4\pi G$ as configuration variable, it follows from the Einstein–Hilbert action that its momentum is $P = \dot{a}/a$: we have the Poisson bracket $\{V, P\} = 1$. In these variables, the Friedmann equation takes the form

[3] A different perspective on the importance of gravitational degrees of freedom is discussed in [11].

$$C := (P^2 + |\Lambda|)V^2 - \frac{1}{12\pi G} p_\phi^2 = 0 \qquad (1)$$

of a constraint rather than an equation of motion. A Wheeler–DeWitt quantization [13] would turn this expression into an operator $\hat{C}$ – for instance in the volume representation where wave functions are of the form $\psi(V, \phi)$ and $\hat{P}$ acts as $-i\hbar\partial/\partial V$ while p_ϕ acts as $-i\hbar\partial/\partial\phi$, with Planck's constant $\hbar$ – and solve the state equation $\hat{C}\psi = 0$. Compared to the Schrödinger equation, time is absent and change would have to be recovered indirectly from the solution space.

What must be absent is time coordinates since they have no role in a quantum theory of gravity, not based on classical space-time manifolds. But other, more physical time parameters may well and should indeed exist. Realizing this is facilitated by eliminating time coordinates already at the classical level, and finding an alternative formulation of classical evolution. To do so, we write equations of motion for V and P with respect to the scalar ϕ. Such equations can be obtained by dividing equations of motion with respect to coordinate time, such as $\mathrm{d}V/\mathrm{d}\phi = \dot{V}/\dot{\phi}$. But any reference to coordinate times can be avoided altogether if we solve the Friedmann equation for the momentum

$$p_\phi = \pm 2\sqrt{3\pi G}\, V \sqrt{P^2 + |\Lambda|} =: H(V, P) \qquad (2)$$

and take $H(V, P)$ as the Hamiltonian for evolution in ϕ. (We will choose the +-sign in what follows, letting ϕ run along with coordinate time.) The Hamiltonian equation of motion $\mathrm{d}O/\mathrm{d}\phi = \{O, H\} = \partial O/\partial V \cdot \partial H/\partial P - \partial O/\partial P \cdot \partial H/\partial V$ for any function O of V and P then equals what we would obtain from dividing coordinate equations of motion.

The case $\Lambda = 0$ is particularly simple. It provides a quadratic Hamiltonian $H \propto |VP|$ and thus constitutes an example of harmonic cosmology [14, 15]. Just as the harmonic oscillator in mechanics, it leads to an exactly solvable quantum system – not just in the sense that solutions can be found in closed form, but with the much stronger property that no quantum back-reaction occurs. The evolution of expectation values is entirely unaffected by changing shapes of a state. In the next section we will see what that entails for dynamics, and how perturbation theory can be used to step from the exactly solvable model to more realistic cases as they are obtained for $\Lambda \neq 0$ or with a non-trivial matter potential.

4 Quantum Dynamics

We now turn to the quantum dynamics of our systems. A quantum system is characterized by the presence of additional, non-classical degrees of freedom which can change independently of the classical variables, given by V and P above. While the latter can be brought in correspondence with expectation values $\langle\hat{V}\rangle$

and $\langle\hat{P}\rangle$ in a quantum state, a whole wave function (or density matrix) contains much more information. Indeed, while classical phase space functions $f(V, P)$ are merely combinations of the canonical coordinates and are fully determined if only a phase space point is specified, products of operators in a quantum system provide independent kinds of information. In general, for instance, $\langle\hat{V}^2\rangle$ can take values irrespective of what the value of $\langle\hat{V}\rangle^2$ is. The difference $(\Delta V)^2 = \langle(\hat{V} - \langle\hat{V}\rangle)^2\rangle$ is a measure for quantum fluctuations, an important quantity in a quantum system. Similarly, all *moments*

$$G^{\overbrace{V\cdots V}^{a}\overbrace{P\cdots P}^{b}} = \langle(\hat{V} - \langle\hat{V}\rangle)^a(\hat{P} - \langle\hat{P}\rangle)^b\rangle_{\mathrm{Weyl}} \tag{3}$$

defined for $a + b \geq 2$ are independent parameters of a (density) state. (The subscript "Weyl" indicates that operator products are ordered totally symmetrically before taking the expectation value.) As discussed in the Appendix, these moments are all dynamical, forming an infinite-dimensional coupled system. Solutions tell us how expectation values of a state behave, but also how the state and its moments evolve. Figure 1 shows the example of a quantum cosmological state during a recollapse, with the spreads changing characteristically.

From these moments we will attempt to form an arrow of time.

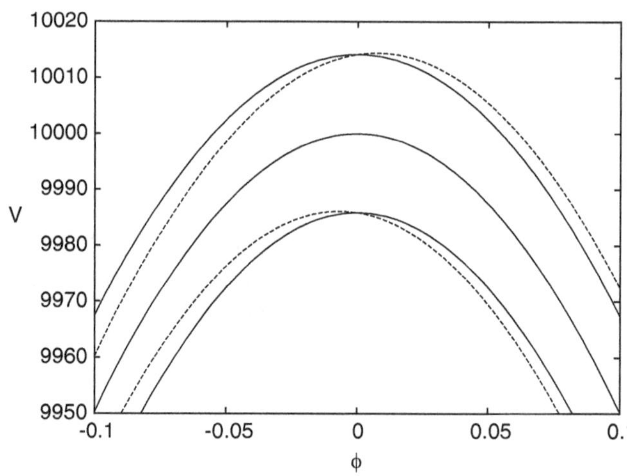

Fig. 1 Dispersing wave at the recollapse of a Friedmann–Robertson–Walker model with $k = 0$ and $\Lambda < 0$. The *top* and *bottom curves* indicate changing spreads ΔV around the central expectation value trajectory of volume V as a function of the scalar ϕ. The latter serves as a measure for time in this free, massless case. For the *solid lines*, the state is unsqueezed, without quantum correlations, at the recollapse point, and fluctuations symmetric around the recollapse result. Non-vanishing correlations (*dashed lines*), on the other hand, lead to non-symmetric fluctuations.

4.1 Monotonicity

When a quantum state changes, its moments change. Just like the expectation values, the moments must satisfy equations of motion which, as derived in the Appendix, follow from the Hamiltonian. Having equations of motion for the second order moments, we can check if any one of them would serve as a good time coordinate [10]. Since quantum states tend to spread out, one may expect fluctuations to have an interpretation of internal time. For G^{PP}, however, this is clearly not the case since its change (16) depends on the sign of curvature $\langle\hat{P}\rangle$. It would decrease in an expanding universe but increase in a collapsing one. Around a recollapse or a bounce, this behavior cannot be monotonic. Similarly, the sign of the rate of change of volume fluctuations G^{VV} is not unique from (18) since neither G^{VP} nor $\langle\hat{P}\rangle$ is required to have a definite sign throughout the history of a universe.

Of more interest for our purposes is the covariance, subject to

$$\frac{\mathrm{d}G^{VP}}{\mathrm{d}\phi} = \frac{3}{2}|\Lambda|\frac{\langle\hat{V}\rangle}{(\langle\hat{P}\rangle^2+|\Lambda|)^{3/2}}G^{PP}$$

from (17). With $G^{PP} = (\Delta P)^2$ required to be positive, the covariance can only grow. Thus, the positivity of fluctuations, or the uncertainty relation in even stronger form, implies a fixed tendency for correlations.

The monotonicity of G^{VP} hints at a possible role in the context of time. In the model considered so far, it certainly does not improve the problem of time since it can anyway be solved trivially by using the scalar (with respect to which G^{VP} now is monotonic). But if we have a look at models with a non-trivial scalar potential $W(\phi) \neq 0$, where ϕ would no longer serve as global time, one can see that G^{VP} is better behaved than just ϕ. In such a case, with a time-dependent potential in the formulation where ϕ plays the role of time, equations can be derived as before provided that the potential is not too large [16, 17]. The classical constraint (still for $\Lambda < 0$) now is

$$\left(P^2+|\Lambda|-\frac{8\pi G}{3}W(\phi)\right)V^2-\frac{1}{12\pi G}p_\phi^2=0$$

and effective equation of motion for the covariance changes to

$$\frac{\mathrm{d}G^{VP}}{\mathrm{d}\phi} = \frac{3}{2}\frac{\langle\hat{V}\rangle(|\Lambda|-8\pi GW(\phi)/3)}{(\langle\hat{P}\rangle^2+|\Lambda|-8\pi GW(\phi)/3)^{3/2}}G^{PP}. \tag{4}$$

For sufficiently small potentials, G^{VP} is still monotonic for wide ranges of evolution. Also here, this refers to monotonicity with respect to ϕ, which now is a good time variable only for finite stretches between turning points in the potential. If we approach a turning point of ϕ, however, the behavior changes. At a turning point, $p_\phi = 0$ and thus $\langle \hat{P} \rangle^2 = -|\Lambda| + 8\pi G W(\phi)/3$ from the constraint. Near a turning point, $|\Lambda| - 8\pi G W(\phi)/3$, appearing in the numerator of (4), thus becomes negative. Even before the turning point of ϕ is reached, G^{VP} according to (4) turns around.

Near a turning point the potential is important and there may be extra terms in the quantum equations of motion. Our analysis at this stage remains incomplete, but it suggests a situation as follows. As a global time variable through periods of oscillation of ϕ, G^{VP} appears no better than the scalar. But it is monotonic in a range around the turning point and can thus be used as local internal time, to which we may transform from ϕ (or other time choices) when a turning point is approached. Thus, it would extend its role of time into a wider region. As a quantum variable without classical analog, this at least suggests that time in a fully relativistic situation can be assisted by quantum aspects.

4.2 *Before the Big Bang*

So far, we have discussed only low-curvature regimes where $P \ll 1$. At larger curvature, new effects from quantum gravity and quantum geometry are expected to take over which are not included in the Wheeler–DeWitt quantization [13] understood up to now. Loop quantum cosmology [18] is one such candidate for an extension, and one of its effects is to provide higher order terms to the Friedmann equation. Its new form then is

$$\frac{\sin^2(\mu P)}{\mu^2} = \frac{1}{12\pi G} \frac{p_\phi^2}{V^2} \tag{5}$$

where μ is a length scale (see, e.g., [19]).

Such higher order terms of P or $\dot{a}$ are expected in quantum gravity if we realize the Friedmann equation as the time-time component of Einstein's tensorial equation. Higher curvature corrections change the action, and thus the Einstein tensor. Correspondingly, the Friedmann equation is amended by higher order terms. (The same reasoning would suggest higher derivative terms, too, which generically are also present. We will, however, be dealing with a solvable model of a free, massless scalar where they are absent [14, 15].) With this analogy, the expansion parameter μ is the same as the one multiplying higher curvature terms, and thus should indeed be dimensionfull. One may think of it as being near the Planck length, which is in fact often assumed. But in loop quantum gravity, it has a dynamical origin related to the discreteness of an underlying quantum gravity state [20, 21]. Generically, μ changes as the universe expands or contracts and cannot always be close to the Planck length. In fact, if it were, other corrections (from inverse scale factor terms [22], based on [23]) would have to be considered as independent quantum corrections, which we avoid here.

The form of the higher order terms, obtained by expanding $\sin(\mu P)$ by powers of P when curvature is small, as well as the length scale μ determining when quantum corrections become important, is not fixed. It may be constrained further by relating such a Hamiltonian to one that is formulated in the full theory, without any symmetry assumptions. But this has currently not been achieved, and so the precise form remains subject to quantization ambiguities. What we discuss in what follows only involves generic qualitative features which depend on some crucial aspects of the loop quantizaton but not on the specific form. As an important effect we include lattice refinement, leading to a possible V-dependence of the parameter μ: the characteristic length scale where discreteness effects happen might depend on the volume and change dynamically [20]. Conceptual [24] as well as phenomenological constraints [25, 26] on the dependence exist, and it is clear that μ cannot be V-independent in all models [27]; but in no case has it been fixed uniquely. A power-law dependence of $\mu \propto V^{\kappa}$ on V, which can realistically describe at least bounded ranges of evolution, can be taken into account by a canonical transformation $P \mapsto V^{\kappa}P$, $V \mapsto V^{1-\kappa}/(1-\kappa)$ which will not change the following results.

Now, the Hamiltonian for ϕ-evolution is not quadratic in V and P even for $k = 0 = \Lambda$, suggesting non-perturbative effects at strong curvature $P \gg 1/\mu$. Fortunately, the system is "resummable" [14]: it is solvable and free of quantum back-reaction if we use the variables V, $J = V\exp(i\mu P)$ instead of canonical ones. These variables satisfy a linear Poisson algebra

$$\{V, J\} = i\mu J, \quad \{V, \bar{J}\} = -i\mu\bar{J}, \quad \{J, \bar{J}\} = -2i\mu V \tag{6}$$

and they provide the basis for solvability even at the dynamical level. In fact, the Hamiltonian for ϕ-evolution, solving (5), is $p_\phi = 2\sqrt{3\pi G}\mathrm{Im}J$ which is linear in the basic variables. Linearity implies that all these relations have direct analogs at the quantum level: $[\hat{V}, \hat{J}] = -\mu\hbar\hat{J}$, $[\hat{V}, \hat{J}^\dagger] = \mu\hbar\hat{J}^\dagger$, and $[\hat{J}, \hat{J}^\dagger] = 2\mu\hbar\hat{V}$ together with the Hamiltonian $\hat{H} = -i\sqrt{3\pi G}(\hat{J} - \hat{J}^\dagger)$. This strong form of solvability allows us to analyze the evolution of a state in precise terms, especially when it approaches the classical singularity.

First, thanks to solvability there is no quantum back-reaction and expectation value equations of motion form a closed set:

$$\frac{\mathrm{d}\langle\hat{V}\rangle}{\mathrm{d}\phi} = \frac{\langle[\hat{V}, \hat{H}]\rangle}{i\hbar} = \sqrt{3\pi G}(\langle\hat{J}\rangle + \langle\hat{J}^\dagger\rangle), \quad \frac{\mathrm{d}\langle\hat{J}\rangle}{\mathrm{d}\phi} = \frac{\langle[\hat{J}, \hat{H}]\rangle}{i\hbar} = 2\sqrt{3\pi G}\langle\hat{V}\rangle. \tag{7}$$

These equations can be combined to $\mathrm{d}^2\langle\hat{V}\rangle/\mathrm{d}\phi^2 = 12\pi G\langle\hat{V}\rangle$, easily integrating to

$$\langle\hat{V}\rangle(\phi) = \alpha\cosh(2\sqrt{3\pi G}\phi) + \beta\sinh(2\sqrt{3\pi G}\phi)$$

with constants of integration α and β to be fixed by initial values. Using (7), we then obtain

$$\mathrm{Re}\,\langle\hat{J}\rangle(\phi) = \frac{1}{2\sqrt{3\pi G}}\frac{\mathrm{d}V}{\mathrm{d}\phi} = \alpha\sinh(2\sqrt{3\pi G}\phi) + \beta\cosh(2\sqrt{3\pi G}\phi).$$

The imaginary part of $\langle\hat{J}\rangle$ is fixed to be

$$\mathrm{Im}\,\langle\hat{J}\rangle(\phi) = \langle\widehat{V\sin(\mu P)}\rangle = \frac{\mu}{2\sqrt{3\pi G}}p_\phi$$

by the preserved ϕ-Hamiltonian, using (5).

The constants of integration α and β determine whether or not $\langle\hat{V}\rangle$ can reach zero, where a singularity would occur. Due to reality conditions, these constants are not arbitrary: classically we have $|J|^2 - V^2 = 0$, which is to be imposed as an operator equation $\hat{J}\hat{J}^\dagger - \hat{V}^2 = 0$ after quantization. (Otherwise the curvature parameter P would not become self-adjoint and physical states obtained by solving the evolution equations would not be correctly normalized.) Since the reality condition is quadratic, it implies

$$0 = \langle\hat{J}\hat{J}^\dagger - \hat{V}^2\rangle = \langle\hat{J}\rangle\langle\hat{J}^\dagger\rangle - \hat{V}^2 + G^{J\bar{J}} - G^{VV} + \mu\hbar\langle\hat{V}\rangle \tag{8}$$

with extra terms from fluctuations. (The last term arises from ordering $\hat{J}\hat{J}^\dagger$ symmetrically.) A state which is semiclassical at a given time has fluctuations of the order $O(\hbar)$, such that the reality condition takes the classical form up to small terms of order $\hbar$. Then, our dynamical solutions must satisfy

$$(\mathrm{Re}\langle\hat{J}\rangle)^2 + (\mathrm{Im}\langle\hat{J}\rangle)^2 - \langle\hat{V}\rangle^2 = -\alpha^2 + \beta^2 + \frac{\mu^2}{12\pi G}p_\phi^2 = O(\hbar)$$

which determines β in terms of α. With $V_{\mathrm{min}} := \mu p_\phi/12\pi G$ and the identity $B\cosh(x + \cosh^{-1}(A/B)) = A\cosh(x) + \sqrt{A^2 - B^2}\sinh(x)$ for arbitrary A and B, the volume is

$$\langle\hat{V}\rangle(\phi) = V_{\mathrm{min}}\cosh(2\sqrt{3\pi G}\phi + \delta) \tag{9}$$

with $\delta = \cosh^{-1}(\alpha/V_{\mathrm{min}})$. This function never becomes zero, proving that the model has a bounce but no singularity. At the bounce point, the density of the scalar field takes the value

$$\rho_{\mathrm{crit}} = \frac{p_\phi^2}{2a^6} = \frac{p_\phi^2}{32\pi^2G^2V_{\mathrm{min}}^2} = \frac{3}{8\pi G\mu^2} \tag{10}$$

which depends on the scale μ but is independent of any initial condition. (The same behavior initially arose from numerical studies [28].)

To evaluate the reality condition, we have used semiclassicality. One might worry that this invalidates conclusions about the bounce, typically expected to occur in a highly quantum regime. However, we had to make assumptions about

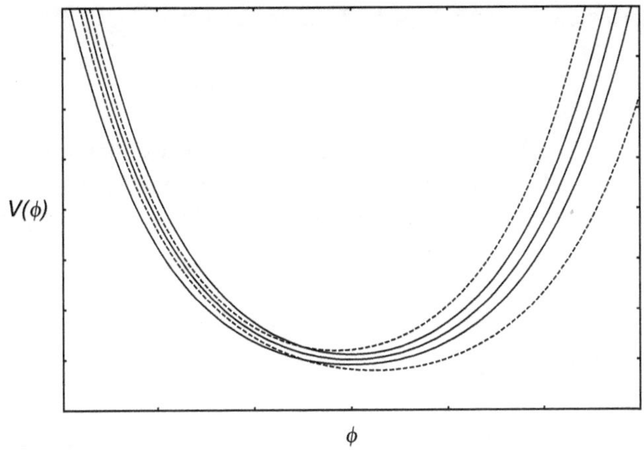

Fig. 2 Dispersing through a bounce. Here, the volume $V(\phi)$ as a function of the scalar, again indicating time, is shown for a bounce rather than a recollapse as in Fig. 1. As before, the *top* and *bottom curves* indicate fluctuations ΔV around the expectation value V – *solid curves* for a state uncorrelated at the bounce, *dashed curves* for a correlated one. Fluctuations "before" the big bang may have been quite different from what they are "afterwards" – see also (11) – to a degree that can be considered forgetful.

semiclassicality only at one time, which can be arbitrarily far away from the bounce. We only need $G^{J\bar{J}} - G^{VV} = O(\hbar)$ throughout, which is at first ensured by an initial condition at large volume. As the state evolves, it may become more quantum. But from equations of motion for the moments it follows that $G^{J\bar{J}} - G^{VV}$ is a constant of motion [15], even if the state spreads, making G^{VV} change. Thus, if this combination is of the order $\hbar$ once, it will remain so. In this solvable model the bounce is realized even for states which may not be semiclassical at the bounce.

The high control in this solvable model persists at the state level. Dispersions as well as squeezing can be followed for general states, as well as specifically for the moments of dynamical coherent states; see Fig. 2 for examples. In principle, we could thus test how covariances evolve and whether they remain monotonic. However, moments with easy access are now those of V and J, not P. Volume fluctuations thus can easily be studied, but the covariance G^{VP} of our earlier interest would, with P related non-linearly to J, be a complicated expression in terms of all the moments involving V and J. Nevertheless, we can find approximate information about the behavior of covariance. Near the bounce, we have $\mu P \sim \pi/2$ for $\sin(\mu P)$ and thus the scalar density to be close to its maximum. This allows us to use the approximation

$$\begin{aligned}\mathrm{Re}\langle\hat{J}\rangle &= \frac{1}{2}\langle\hat{V}\widehat{e^{i\mu P}} + \widehat{e^{-i\mu P}}\hat{V}\rangle \sim \frac{1}{2}\langle e^{i\pi/2}\hat{V}i(\mu\hat{P} - \pi/2) - e^{-i\pi/2}i(\mu\hat{P} - \pi/2)\hat{V}\rangle \\ &= -\frac{\mu}{2}\langle\hat{V}\hat{P} + \hat{P}\hat{V}\rangle + \frac{\pi}{2}\langle\hat{V}\rangle \sim -\mu G^{VP}\end{aligned}$$

by Taylor expansion around $\mu P \sim \pi/2$. Noting that

$$\mathrm{Re}\langle \hat{J} \rangle = V_{\min} \sinh(2\sqrt{3\pi G}\phi + \delta)$$

from (7) and (9) is monotonic in ϕ, we are led to suggest that also the covariance on the right hand side is monotonic. Combining all conclusions, it will thus be a good measure for time through several cosmological phases, including recollapses and bounces.

5 Beyond Exactitude

So far, we have considered a free, exactly solvable model to shine some light on the universe at small volume. Such models rarely give the full picture of a physical situation they may be applied to. There are several additional ingredients to be required for a physically reliable analysis of a whole universe through and before the big bang, most importantly inhomogeneous configurations. No general description of inhomogeneities is available around bounce regimes in loop quantum cosmology, not even in perturbative form.

A crucial issue is that of the consistency of higher order terms, such as those appearing in (5), in a context which is no longer homogeneous. Then, the full anomaly issue strikes and modifications to the classical constraints are highly restricted: it is not easy to implement quantum corrections while still maintaining the same level of general covariance as it is realized classically. If covariance transformations are broken, the theory will be anomalous and inconsistent; such transformations could only be deformed by quantum corrections but must remain present in the same number. (Effective actions starting from quantum corrected isotropic equations have been determined [29–31]. But trying to embed a finite-dimensional model in a fully inhomogeneous system is highly ambiguous, and so quantum corrections for inhomogeneities remain unknown in the presence of higher order corrections such as (5); examples do, however, exist for special modes [32,33] or other effects of loop quantum gravity [34–38].)

Consistency issues arise due to general covariance, which implies that one is dealing with a system of constraints, or an overdetermined set of equations. While there is only one, trivially consistent constraint (5) in isotropic models, several independent ones exist when geometries become inhomogeneous. Their algebra under Poisson brackets obeys certain conditions for the system to be well-defined, which must also be realized for the quantum representation. A possibility to sidestep the quantization of constraints is reduced phase space quantization, where one tries to find the classical solution space to all constraints and quantizes it. The usual problems are that (1) constraints may be difficult to solve completely and (2) the solution space may be of complicated structure, for instance in its topological properties, and thus be difficult to quantize in its own right.

In the context of perturbative inhomogeneities, the first problem does not arise at least at the linear level since all gauge-invariant perturbations can easily be written down; see e.g. [39–45]. For linear perturbations, moreover, topological properties of solution spaces mostly disappear such that a reduced quantization here may be viable [46, 47]. Alas, it cannot present a full theory if it is simply added on to the bouncing background as treated so far, which was by the Dirac rather than the reduced phase space procedure. One may deal with the background dynamics also by reduced phase space techniques [48, 49], but that would work easily only for a free, massless scalar trivializing the problem of time. By the Dirac procedure, on the other hand, the theory can be formulated for general interacting scalars [50], even though it may be solved easily only in free scalar cases or perturbations around those.

In a reduced phase space quantization of perturbative inhomogeneities, no fully defined theory would be available. This may be acceptable if it can be seen as a valid approximation to some other full theory, but this is not the case. In fact, in systems not involving the bounce, where consistent quantizations of perturbative inhomogeneities in loop quantum gravity are available [34], one can see that a reduced phase space quantization would overlook crucial effects. As shown in [35], quantum corrections can induce effective anisotropic stress terms even in systems which classically have no anisotropic stress. A reduced phase space formulation based on the classical identities between gauge-invariant quantities could not see this new quantum effect, and thus must remain incomplete. Similarly, gauge-fixed treatments (as used e.g. in [51, 52] for recent examples) often hide crucial quantum properties. (Also the more general reduced phase space formulation of [44, 45] is subject to these remarks. Moreover, even in this reduced context, consistency conditions remain which are yet to be implemented in a possible quantization. While valuable, these formulations so far do not suffice to see how an inhomogeneous universe may evolve through a bounce.)

Such a situation makes the task of developing cosmological scenarios based on bounces difficult. But some indications can nonetheless be derived from models if they are understood for the local behavior of a patch of space-time near the moment of its smallest size. How different patches connect may be impossible to say in the absence of a fully inhomogeneous description, but the evolution of a single patch may still carry some surprises. Concrete properties, such as the density when a patch bounces, may easily change or go away when a sufficiently general situation is considered. But in addition to such positive, affirmative properties there are negative ones which tell us about limitations of what can be said for early stages of the patch. Negative statements of this form are much more reliable, for if knowledge of something is constrained in a simple model, it is unlikely to become better known in a general situation.

There is such a negative property which, rather surprisingly, shows up even in the exactly solvable model [53]. It is not about classical variables, or the expectation values, but rather about quantum fluctuations or other moments. As before, we can derive equations of motion for all the moments, say of second order, forming a closed set of equations. There are several independent second order moments and their equations, with correspondingly many initial values to be chosen for a state. One can cut down the number of parameters by selecting dynamical coherent states: those that saturate the uncertainty relations at all times. The calculations are somewhat lengthy but can be completed [54], with the result that volume fluctuations at early and late times are related by

$$\Delta := \left| \lim_{\phi\to-\infty} \frac{G^{VV}}{\langle\hat{V}\rangle^2} - \lim_{\phi\to\infty} \frac{G^{VV}}{\langle\hat{V}\rangle^2} \right| \tag{11}$$

$$= 4\frac{H}{V_{\mathrm{min}}} \sqrt{\left(1 - \frac{H^2}{V_{\mathrm{min}}^2} + \frac{1}{4}\frac{\hbar^2}{V_{\mathrm{min}}^2}\right)\frac{(\Delta H)^2}{V_{\mathrm{min}}^2} - \frac{1}{4}\frac{\hbar^2}{V_{\mathrm{min}}^2} + \left(\frac{H^2}{V_{\mathrm{min}}^2} - 1\right)\frac{(\Delta H)^4}{V_{\mathrm{min}}^4}}$$

where H is the expectation value of the ϕ-Hamiltonian and ΔH its fluctuation. This parameterizes the behavior for all dynamical coherent states.

Of particular interest is the behavior when H is large, which means that one would use the model for a patch containing a large amount of matter. As shown in Fig. 3, the asymmetry in such a case depends very sensitively on the parameters, for instance the ratio V_{min}/H. Moreover, its value can differ significantly from one; fluctuations of the state by no means have to remain unchanged when phases before and after the bounce are considered. There is a degree of cosmic forgetfulness [53]:

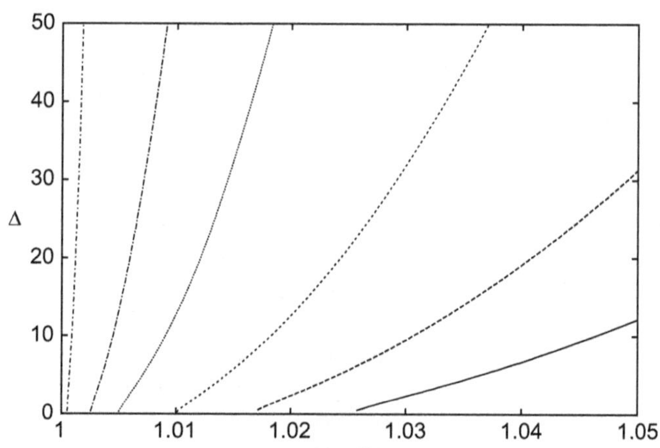

Fig. 3 Sensitivity: The asymmetry Δ of volume fluctuations from (11), depending on the ratio V_{min}/H with H the value of the scalar momentum as a measure for the amount of matter. Different curves correspond to various values of H, growing to the left. Thus, the steep leftmost curves are obtained for a universe with a large amount of matter, the more realistic scenario within the simple solvable models used here. The asymmetry (11) depends very sensitively on the initial values that determine V_{min}/H; unrealistically sensitive measurements would be required at one side of the bounce to determine the volume fluctuations at the other side. It is practically impossible to recover the complete state due to this cosmic forgetfulness [54].

due to the high sensitivity it is practically impossible to recover the full state before the bounce from its properties after the bounce.

6 An Arrow of Moments

In our discussion of the covariance, the big bang, resolved to a bounce, did not appear special in any way regarding the direction of time. It did not suggest a turn-around in the rate of change of the covariance. Had it done so, it would have led us to conclude that G^{VP} cannot serve as a good time in that phase, rather than suggesting a flip of the arrow of time.

The role of moments concerning the arrow of time is more subtle. We will first reformulate the usual context to see how it may be related to evolving quantum states. One often says that what distinguishes the past from the future is that we remember the former and try to predict the latter. It may be more honest to define the past as what we can (and typically do) forget. For human behavior, one of the most important and most annoying consequences of the arrow of time is indeed forgetfulness. In a more general sense, this is true also for thermodynamical systems, although it may not be so clear whether this is really annoying. A thermodynamical system evolving toward equilibrium forgets any sense of being special as it might have been encoded in its initial configuration. In quantum cosmology, even the whole universe has a case of cosmic forgetfulness which one may relate to the arrow of time.[4]

To illustrate this, we return to the resummed solvable model of loop quantum gravity. Now considering its own moments for V and J, we can look for all choices giving rise to dynamical coherent states: evolving states which saturate the uncertainty relations at all times. Such states provide the best control one may have on a quantum system, and thus highlight when anything becomes inaccessible – for instance by being forgotten. As already described, this is exactly what happens. Although one could not easily use the solvable model to draw strong conclusions about the universe before the big bang, what it tells us about limitations has to be taken seriously.[5]

The kind of cosmic forgetfulness realized in this model provides an orientation of time, telling us not only which of the properties before the big bang can be forgotten,

---

[4]In thermodynamics, coarse-graining plays an important role. Cosmic forgetfulness may be interpreted as forcing us to coarse-grain over many of the quantum variables. One should also note that cosmic forgetfulness is much stronger than decoherence (see e.g. [3]) since it appears even in exactly solvable models. It takes into account the specific dynamics of loop quantum cosmology, rather than being a generic property of quantum systems with many degrees of freedom.

[5]Cosmic forgetfulness has been perceived as a challenge, heroically taken up in [55] by deriving bounds alternative to (11) for semiclassical states. However, those bounds are much weaker, allowing changes in the fluctuations by several orders of magnitude [56]. (Also this renewed challenge has been taken up in [57], though less heroically so.)

but also what direction "before the big bang" is. An observer after the bounce would be unable to reconstruct the full state before the bounce, but could easily predict the future development toward larger volume. This arrow agrees with the standard notion.

Now asking how an observer before the big bang would experience the same situation, the answer is also clear: such an observer would be unable to determine the precise state at larger values of ϕ beyond the bounce, but could easily extrapolate the state to smaller values of ϕ. The state at smaller values of ϕ can be predicted, while the state at large values of ϕ is forgotten once the bounce is penetrated. Since one cannot forget the future, such an observer must be attributed a reversed arrow of time, pointing toward smaller ϕ. At the bounce, two arrows would emerge pointing in opposite directions as far as ϕ is concerned. In this sense, the model resembles [58–61].

While degrees of freedom propagating in a bouncing universe still have to be understood much better, indications do exist that what appears as a simple bounce in homogeneous models may have to be interpreted rather differently when degrees of freedom other than the purely classical homogeneous ones are considered. Here, this has been discussed for homogeneous quantum degrees of freedom; inhomogeneities will be the next crucial and decisive step.[6]

7 Conclusions

"If he had smiled why would he have smiled? To reflect that each one who enters imagines himself to be the first to enter whereas he is always the last term of a preceding series even if the first term of a succeeding one, each imagining himself to be first, last, only and alone, whereas he is neither first nor last nor only nor alone in a series originating in and repeated to infinity."[7] This describes the thoughts of Leopold Bloom after a long eventful day. Will we be led to similar thoughts after a long eventful journey in quantum gravity?

If we cannot reconstruct the entire past, we may as well forget about it. The part of the universe we see would appear to have originated with its big bang, even though a theoretical formulation, but only the theoretical formulation, may contain a pre-history. Two questions should immediately be asked: Would this be testable? And why would we not apply Occam's razor to the pre-history? Clearly, we could not directly test whether there is a part of the history of the universe that is inaccessible. But we may attempt to access it and, if we succeed, falsify the claim; this makes it scientifically viable as a hypothesis. More importantly, the underlying scenario would have further implications for the structures we see after the big bang. Then, we would have an option to test such a model indirectly.

---

[6] Numerical indications for a similarly sensitive behavior of inhomogeneities [62] already exist from Gowdy models with a loop-quantized homogeneous background [63].

[7] James Joyce: Ulysses.

Why do we then consider the pre-history as part of the mathematical modelling? Also this has its justification. Describing a true physical beginning of the universe, where nothing would turn into something, has proved to be challenging. Pretending that there was something before the big bang and describing it by deterministic but forgetful equations may be the best solution to deal with a beginningless beginning, even though we may not be able to use those equations to fully reconstruct the past.

Taking the simplest models of loop quantum cosmology at face value is often seen as suggesting the big bang transition to be viewed as a smooth bounce, as one further element not just in a long history of the universe itself but also in a long history of bouncing cosmological models [64]. Some indications, however, suggest otherwise. The bloomy scenario of loop quantum cosmology may well be this: a universe whose time-reversed pre-history we cannot access but which we grasp in the form of initial conditions it provides for our accessible part; a pseudo-beginning [65]; an orphan universe, shown the rear-end by whatever preceded (and possibly created) it.

Acknowledgements This work was partially supported by NSF grant PHY0748336 and a grant from the Foundational Questions Institute (FQXi).

Appendix: A Momentous Formulation of Quantum Mechanics

Quantum dynamics can usefully be described in terms of the moments (3) of a state. Taken together, they form an infinite-dimensional phase space which can be used to describe the quantum system. At order $a+b=2$ we have the fluctuations $G^{2,0}=(\Delta V)^2$ and $G^{0,2}=(\Delta P)^2$ as well as the covariance $G^{1,1}=\frac{1}{2}\langle\hat{V}\hat{P}+\hat{P}\hat{V}\rangle-\langle\hat{V}\rangle\langle\hat{P}\rangle$. While independent variables, the moments cannot be chosen arbitrarily. They are subject to constraints, most importantly the uncertainty relation

$$G^{VV}G^{PP}-(G^{VP})^2\geq\frac{\hbar^2}{4}.$$

Poisson brackets between the moments can be computed using the general identity

$$\{\langle\hat{A}\rangle,\langle\hat{B}\rangle\}=\frac{\langle[\hat{A},\hat{B}]\rangle}{i\hbar} \tag{12}$$

as well as linearity and the Leibniz rule. This immediately gives $\{\langle\hat{V}\rangle,\langle\hat{P}\rangle\}=1$ and, e.g., $\{G^{VV},G^{PP}\}=4G^{VP}$. (See [66, 67] for further details.)

The moments allow a convenient description of quantum evolution without having to take the usual detour of solving for states first, followed by computing expectation values. Instead, expectation values obey the general evolution law

$$\frac{d\langle \hat{O} \rangle}{d\phi} = \frac{\langle [\hat{O}, \hat{H}] \rangle}{i\hbar}$$

which can be used to derive a coupled set of equations of motion for expectation values together with the moments. For non-polynomial $\hat{H}$, it may be difficult to compute the commutator, followed by taking an expectation value. Semiclassical equations can more easily be obtained in an expansion by moments, which is formally analogous to background-field expansion around expectation values. We write [66]

$$\langle H(\hat{V}, \hat{P}) \rangle = \langle H(\langle \hat{V} \rangle + (\hat{V} - \langle \hat{V} \rangle), \langle \hat{P} \rangle + (\hat{P} - \langle \hat{P} \rangle)) \rangle \tag{13}$$

$$= H(\langle \hat{V} \rangle, \langle \hat{P} \rangle) + \sum_{a,b:a+b\geq 2} \frac{1}{a!b!} \frac{\partial^{a+b} H(\langle \hat{V} \rangle, \langle \hat{P} \rangle)}{\partial \langle \hat{V} \rangle^a \partial \langle \hat{P} \rangle^b} G^{a,b} \tag{14}$$

and use this in

$$\frac{\langle [\hat{O}, \hat{H}] \rangle}{i\hbar} = \{ \langle \hat{O} \rangle, \langle H(\hat{V}, \hat{P}) \rangle \}. \tag{15}$$

Poisson relations between the moments then provide all equations of motion order by order in the moments.

For the cosmological systems with Hamiltonian (2) introduced before, we have

$$\frac{d\langle \hat{V} \rangle}{d\phi} = \frac{3}{2} \frac{\langle \hat{V} \rangle \langle \hat{P} \rangle}{\sqrt{\langle \hat{P} \rangle^2 + |\Lambda|}} - \frac{9}{4} |\Lambda| \frac{\langle \hat{V} \rangle \langle \hat{P} \rangle}{(\langle \hat{P} \rangle^2 + |\Lambda|)^{5/2}} G^{PP} + \frac{3}{2} |\Lambda| \frac{G^{VP}}{(\langle \hat{P} \rangle^2 + |\Lambda|)^{3/2}} + \cdots$$

$$\frac{d\langle \hat{P} \rangle}{d\phi} = -\frac{3}{2} \sqrt{\langle \hat{P} \rangle^2 + |\Lambda|} - \frac{3}{4} |\Lambda| \frac{G^{PP}}{(\langle \hat{P} \rangle^2 + |\Lambda|)^{3/2}} + \cdots$$

expanded by the moments (kept here to second order only). This is accompanied by the evolution of moments

$$\frac{dG^{PP}}{d\phi} = -3 \frac{\langle \hat{P} \rangle}{\sqrt{\langle \hat{P} \rangle^2 + |\Lambda|}} G^{PP} + \cdots \tag{16}$$

$$\frac{dG^{VP}}{d\phi} = \frac{3}{2} |\Lambda| \frac{\langle \hat{V} \rangle}{(\langle \hat{P} \rangle^2 + |\Lambda|)^{3/2}} G^{PP} + \cdots \tag{17}$$

$$\frac{dG^{VV}}{d\phi} = 3|\Lambda| \frac{\langle \hat{V} \rangle}{(\langle \hat{P} \rangle^2 + |\Lambda|)^{3/2}} G^{VP} + 3 \frac{\langle \hat{P} \rangle}{\sqrt{\langle \hat{P} \rangle^2 + |\Lambda|}} G^{VV} + \cdots. \tag{18}$$

Solving or analyzing this coupled set of equations would tell us how the state changes its shape by the evolving moments, and how this back-reacts on the motion of expectation values. In some regimes it is possible to summarize the effect of all moments in an effective potential for expectation values depending only on the classical variables. But in general, higher-dimensional effective systems including the moments as independent variables are required.

References

1. Ch. Berger, L. Sehgal, CP violation and arrows of time evolution of a neutral K or B meson from an incoherent to a coherent state. Phys. Rev. D **76**, 036003 (2007) [arXiv:0704.1232]
2. K.V. Kuchař, Time and interpretations of quantum gravity, in *Proceedings of the 4th Canadian Conference on General Relativity and Relativistic Astrophysics*, ed. by G. Kunstatter, D. E. Vincent, J.G. Williams, (World Scientific, Singapore, 1992)
3. H.D. Zeh, Open questions regarding the arrow of time, in this volume. (Springer, Berlin) [arXiv:0908.3780]
4. M. Gasperini, M. Giovannini, Quantum squeezing and cosmological entropy production. Class. Quantum Grav. **10**, L133–L136 (1993)
5. M. Kruczenski, L.E. Oxman, M. Zaldarriaga, Large squeezing behaviour of cosmological entropy generation. Class. Quantum Grav. **11**, 2317–2329 (1994)
6. D. Koks, A. Matacz, B.L. Hu, Entropy and uncertainty of squeezed quantum open systems. Phys. Rev. D **55**, 5917–5935 (1997) (Erratum: [7])
7. D. Koks, A. Matacz, B.L. Hu, Phys. Rev. D **56**, 5281 (1997)
8. C. Kiefer, D. Polarski, A.A. Starobinsky, Entropy of gravitons produced in the early universe. Phys. Rev. D **62**, 043518 (2000)
9. M. Bojowald, B. Sandhöfer, A. Skirzewski, A. Tsobanjan, Effective constraints for quantum systems. Rev. Math. Phys. **21**, 111–154 (2009) [arXiv:0804.3365]
10. M. Bojowald, R. Tavakol, Recollapsing quantum cosmologies and the question of entropy. Phys. Rev. D **78**, 023515 (2008) [arXiv:0803.4484]
11. L. Mersini-Houghton, Notes on Time's Enigma [arXiv:0909.2330]
12. W. Kaminski, J. Lewandowski, The flat FRW model in LQC: The self-adjointness. Class. Quant. Grav. **25**, 035001 (2008) [arXiv:0709.3120]
13. D.L. Wiltshire, An introduction to quantum cosmology, in *Cosmology: The Physics of the Universe*, ed. by B. Robson, N. Visvanathan, W.S. Woolcock (World Scientific, Singapore, 1996) pp. 473–531 [gr-qc/0101003]
14. M. Bojowald, Large scale effective theory for cosmological bounces. Phys. Rev. D **75**, 081301(R) (2007) [gr-qc/0608100]
15. M. Bojowald, Dynamical coherent states and physical solutions of quantum cosmological bounces. Phys. Rev. D **75**, 123512 (2007) [gr-qc/0703144]
16. M. Bojowald, H. Hernández, A. Skirzewski, Effective equations for isotropic quantum cosmology including matter. Phys. Rev. D **76**, 063511 (2007) [arXiv:0706.1057]
17. M. Bojowald, A. Tsobanjan, Effective constraints for relativistic quantum systems. Phys. Rev. D **80**, 125008 (2009) [arXiv:0906.1772]
18. M. Bojowald, Loop quantum cosmology. Living Rev. Relativity **11**, 4 (2nd July 2008) [gr-qc/0601085] http://www.livingreviews.org/lrr-2008-4
19. K. Banerjee, G. Date, Discreteness corrections to the effective hamiltonian of isotropic loop quantum cosmology. Class. Quant. Grav. **22**, 2017–2033 (2005) [gr-qc/0501102]
20. M. Bojowald, Loop quantum cosmology and inhomogeneities. Gen. Rel. Grav. **38**, 1771–1795 (2006) [gr-qc/0609034]

21. M. Bojowald, The dark side of a patchwork universe. Gen. Rel. Grav. **40**, 639–660 (2008) [arXiv:0705.4398]
22. M. Bojowald, Inverse scale factor in isotropic quantum geometry. Phys. Rev. D **64**, 084018 (2001) [gr-qc/0105067]
23. T. Thiemann, QSD V: Quantum gravity as the natural regulator of matter quantum field theories. Class. Quantum Grav. **15**, 1281–1314 (1998) [gr-qc/9705019]
24. M. Bojowald, D. Cartin, G. Khanna, Lattice refining loop quantum cosmology, anisotropic models and stability. Phys. Rev. D **76**, 064018 (2007) [arXiv:0704.1137]
25. W. Nelson, M. Sakellariadou, Lattice refining loop quantum cosmology and inflation. Phys. Rev. D **76**, 044015 (2007) [arXiv:0706.0179]
26. W. Nelson, M. Sakellariadou, Lattice refining LQC and the matter hamiltonian. Phys. Rev. D **76**, 104003 (2007) [arXiv:0707.0588]
27. M. Bojowald, Consistent loop quantum cosmology. Class. Quantum Grav. **26**, 075020 (2009) [arXiv:0811.4129]
28. A. Ashtekar, T. Pawlowski, P. Singh, Quantum nature of the big bang: Improved dynamics. Phys. Rev. D **74**, 084003 (2006) [gr-qc/0607039]
29. G.J. Olmo, P. Singh, Covariant effective action for loop quantum cosmology a la Palatini. J. Cosmology Astropart. Phys. **0901**, 030 (2009) [arXiv:0806.2783]
30. G. Date, S. Sengupta, Effective actions from loop quantum cosmology: Correspondence with higher curvature gravity. Class. Quant. Grav. **26**, 105002 (2009) [arXiv:0811.4023]
31. T.P. Sotiriou, Covariant effective action for loop quantum cosmology from order reduction. Phys. Rev. D **79**, 044035 (2009) [arXiv:0811.1799]
32. M. Bojowald, G. Hossain, Cosmological vector modes and quantum gravity effects. Class. Quantum Grav. **24**, 4801–4816 (2007) [arXiv:0709.0872]
33. M. Bojowald, G. Hossain, Quantum gravity corrections to gravitational wave dispersion. Phys. Rev. D **77**, 023508 (2008) [arXiv:0709.2365]
34. M. Bojowald, G. Hossain, M. Kagan, S. Shankaranarayanan, Anomaly freedom in perturbative loop quantum gravity. Phys. Rev. D **78**, 063547 (2008) [arXiv:0806.3929]
35. M. Bojowald, G. Hossain, M. Kagan, S. Shankaranarayanan, Gauge invariant cosmological perturbation equations with corrections from loop quantum gravity. Phys. Rev. D **79**, 043505 (2009) [arXiv:0811.1572]
36. M. Bojowald, T. Harada, R. Tibrewala, Lemaitre-Tolman-Bondi collapse from the perspective of loop quantum gravity. Phys. Rev. D **78**, 064057 (2008) [arXiv:0806.2593]
37. M. Bojowald, J.D. Reyes, Dilaton gravity, poisson sigma models and loop quantum gravity. Class. Quantum Grav. **26**, 035018 (2009) [arXiv:0810.5119]
38. M. Bojowald, J.D. Reyes, R. Tibrewala, Non-marginal LTB-like models with inverse triad corrections from loop quantum gravity. Phys. Rev. D **80**, 084002 (2009) [arXiv:0906.4767]
39. J.M. Bardeen, Gauge-invariant cosmological perturbations. Phys. Rev. D **22**, 1882–1905 (1980)
40. D. Langlois, Hamiltonian formalism and gauge invariance for linear perturbations in inflation. Class. Quant. Grav. **11**, 389–407 (1994)
41. E.J.C. Pinho, N. Pinto-Neto, Scalar and vector perturbations in quantum cosmological backgrounds. Phys. Rev. D **76**, 023506 (2007) [arXiv:hep-th/0610192]
42. B. Dittrich, J. Tambornino, A perturbative approach to Dirac observables and their space-time algebra. Class. Quant. Grav. **24**, 757–784 (2007) [gr-qc/0610060]
43. B. Dittrich, J. Tambornino, Gauge invariant perturbations around symmetry reduced sectors of general relativity: Applications to cosmology, Class. Quantum Grav. **24**, 4543–4585 (2007) [gr-qc/0702093]
44. K. Giesel, S. Hofmann, T. Thiemann, O. Winkler, Manifestly Gauge-Invariant General Relativistic Perturbation Theory: I. Foundations. Class. Quant. Grav. **27**, 055005 (2010) [arXiv:0711.0115]
45. K. Giesel, S. Hofmann, T. Thiemann, O. Winkler, Manifestly Gauge-Invariant General Relativistic Perturbation Theory: II. FRW Background and First Order. Class. Quant. Grav. **27**, 055006 (2010) [arXiv:0711.0117]

46. F.T. Falciano, N. Pinto-Neto, Scalar perturbations in scalar field Quantum cosmology. Phys. Rev. D **79**, 023507 (2009) [arXiv:0810.3542]
47. J. Puchta, Master thesis, Warsaw University
48. P. Dzierzak, P. Malkiewicz, W. Piechocki, Turning Big Bang into Big Bounce: I. Classical Dynamics. Phys. Rev. D **80**, 104001 (2009) [arXiv:0907.3436]
49. P. Malkiewicz, W. Piechocki, Turning big bang into big bounce: Quantum dynamics. Class. Quant. Grav. **27**, 225018 (2010) [arXiv:0908.4029]
50. W.F. Blyth, C.J. Isham, Quantization of a Friedmann universe filled with a scalar field. Phys. Rev. D **11**, 768–778 (1975)
51. M. Artymowski, Z. Lalak, L. Szulc, Loop quantum cosmology corrections to inflationary models. J. Cosmology Astropart. Phys. **0901**, 004 (2009) [arXiv:0807.0160]
52. J. Mielczarek, The Observational Implications of Loop Quantum Cosmology. Phys. Rev. D **81**, 063503 (2010) [arXiv:0908.4329]
53. M. Bojowald, What happened before the big bang? Nat. Phys. **3**, 523–525 (2007)
54. M. Bojowald, Harmonic cosmology: How much can we know about a universe before the big bang? Proc. Roy. Soc. A **464**, 2135–2150 (2008) [arXiv:0710.4919]
55. A. Corichi, P. Singh, Quantum bounce and cosmic recall. Phys. Rev. Lett. **100**, 161302 (2008) [arXiv:0710.4543]
56. M. Bojowald, Comment on Quantum bounce and cosmic recall. Phys. Rev. Lett. **101**, 209001 (2008) [arXiv:0811.2790]
57. A. Corichi, P. Singh, Reply to 'Comment on Quantum Bounce and Cosmic Recall, Phys. Rev. Lett. **101**, 209002 (2008) [arXiv:0811.2983]
58. C. Kiefer, H.D. Zeh, Arrow of time in a recollapsing quantum universe. Phys. Rev. D **51**, 4145–4153 (1995) [gr-qc/9402036]
59. A. Aguirre, S. Gratton, Steady-state eternal inflation. Phys. Rev. D **65**, 083507 (2002) [astro-ph/0111191]
60. A. Aguirre, S. Gratton, Inflation without a beginning: A null boundary proposal. Phys. Rev. D **67**, 083515 (2003) [gr-qc/0301042]
61. S.M. Carroll, J. Chen, Spontaneous Inflation and the Origin of the Arrow of Time (2004) [hep-th/0410270]
62. D. Brizuela, G.A. Mena Marugan, T. Pawlowski, Big Bounce and inhomogeneities (2009). Class. Quant. Grav. **27**, 052001 (2010) [arXiv:0902.0697]
63. M. Martín-Benito, L.J. Garay, G.A. Mena Marugán, Hybrid quantum gowdy cosmology: Combining loop and fock quantizations. Phys. Rev. D **78**, 083516 (2008) [arXiv:0804.1098]
64. M. Novello, S.E.P. Bergliaffa, Bouncing cosmologies. Phys. Rep. **463**, 127–213 (2008)
65. R. Vaas, Time before Time - Classifications of universes in contemporary cosmology, and how to avoid the antinomy of the beginning and eternity of the world (2004) [physics/0408111]; see also R. Vaas, Time after time - big bang cosmology and the arrow of time (2012), this volume
66. M. Bojowald, A. Skirzewski, Effective equations of motion for quantum systems. Rev. Math. Phys. **18**, 713–745 (2006) [math-ph/0511043]
67. M. Bojowald, A. Skirzewski, Quantum gravity and higher curvature actions. Int. J. Geom. Meth. Mod. Phys. **4**, 25–52 (2007) [hep-th/0606232]; in *Proceedings of Current Mathematical Topics in Gravitation and Cosmology (42nd Karpacz Winter School of Theoretical Physics)*, ed. by A. Borowiec, M. Francaviglia

Can the Arrow of Time Be Understood from Quantum Cosmology?

Claus Kiefer

Abstract I address the question whether the origin of the observed arrow of time can be derived from quantum cosmology. After a general discussion of entropy in cosmology and some numerical estimates, I give a brief introduction into quantum geometrodynamics and argue that this may provide a sufficient framework for studying this question. I then show that a natural boundary condition of low initial entropy can be imposed on the universal wave function. The arrow of time is then correlated with the size of the Universe and emerges from an increasing amount of decoherence due to entanglement with unobserved degrees of freedom. Remarks are also made concerning the arrow of time in multiverse pictures and scenarios motivated by dark energy.

1 Introduction

The fundamental laws of physics, as they are presently known, are mostly invariant with respect to a reversal of time: to every solution there exists an equally viable solution in which t is replaced by $-t$. The only exceptions are some processes described by the weak interaction, but these cases can also be subsumed under time-reversal invariance in the broad sense because its violation there can directly be compensated by an application of a unitary CP-transformation; the latter is possible because the combination of charge conjugation (C), parity transformation (P), and time reversal (T) is conserved (CPT-theorem).

Despite this fundamental invariance, most classes of phenomena observed in Nature distinguish a specific direction of time. These are the famous "arrows of time" which are discussed at length in [1], see also [2] and other contributions

C. Kiefer (✉)
Institut für Theoretische Physik, Universität zu Köln, Zülpicher Straße 77, 50937 Köln, Germany
e-mail: kiefer@thp.uni-koeln.de

L. Mersini-Houghton and R. Vaas (eds.), *The Arrows of Time*, Fundamental Theories of Physics 172, DOI 10.1007/978-3-642-23259-6_10,

to this volume. This observed discrepancy between the symmetric laws and the asymmetric facts does not constitute an inconsistency; asymmetric phenomena are compatible with symmetric laws, and most solutions of the fundamental equations are not symmetric by themselves. What is peculiar is the fact that the time direction of the phenomena is always the same.

Usually one distinguishes between various manifestations of the arrow of time [1]. The electrodynamic arrow expresses the preference of the retarded over the advanced solutions. The thermodynamic arrow is given by the non-decrease of entropy for closed systems, as expressed by the Second Law of thermodynamics. The electrodynamic arrow can, in fact, be derived from the Second Law by using the thermodynamical properties of absorbers. The quantum-mechanical arrow expresses a direction through the measurement process or, in the Everett picture, the branching of the wave function. A central role is there played by decoherence – the irreversible and unavoidable emergence of classical properties through interaction with the environment. Finally, gravitational systems exhibit a preferred direction either through gravitational collapse or through the expansion of the Universe. The question raised by the presence of all these arrows is whether a common *master* arrow of time is behind all of them.

One might wonder whether the arrow of time points into different directions for different subsystems of the Universe, for example, for different galaxy clusters. Using arguments that can be traced back to Emile Borel in the 1920s, one can recognize that there is no strict isolation of subsystems and that therefore all arrows in the Universe must point into the *same* direction. This suggests that the master arrow of time may be found in cosmology.

Can one explain the presence of these distinguished time directions in the framework of physics? Is there a master arrow of time and, if yes, where does it come from? One might speculate that a new, hitherto unknown, fundamental law exists, which is asymmetric in time. Models of wave-function collapse are explicit examples for such new laws. However, no empirical evidence exists for them. The alternative to such a speculation is the presence of a distinguished cosmic boundary condition of low entropy at or near the Big Bang. One would then expect that the entropy of the Universe increases in the direction of increasing cosmic size. The question remains, however, where such a boundary condition could come from.

As indicated by the singularity theorems of general relativity, a consistent description of the Big Bang may require a new framework such as quantum gravity. The question then arises whether the origin of the arrow of time can be understood there. This is the topic of my essay. I shall start in the next section by making more precise the arguments in favour of a cosmic boundary condition of low entropy. I shall then present a framework of quantum gravity in which the above question can be addressed – quantum geometrodynamics, the direct quantization of Einstein's theory of general relativity. In the last section I shall then present how, in fact, the origin of irreversibility could be understood from quantum cosmology. I shall also speculate there about possible quantum effects and the fate of the Second Law in the future of our Universe. The Appendix contains numerical estimates concerning the entropy and the maximal entropy of our Universe.

2 Entropy and Cosmology

Already Ludwig Boltzmann had speculated that the Second Law has its origin in a gigantic fluctuation in the Universe. His picture was that the Universe is eternally existing and at its maximal entropy most of the time, but that at very rare occasions (which, of course, can happen in an eternal Universe) the entropy fluctuates to a very low value from which it will then increase; this would then enable our existence and lead to the arrow of time that we observe. The weak point in this argument was disclosed in the 1930s by Carl Friedrich von Weizsäcker: if one takes into account the possibility of entropy fluctuations, a fluctuation that produces at once the world that we observe including our existence and our memories, although by itself extremely unlikely, would still be much more probable than Boltzmann's fluctuation which has to create the whole history of the world in addition to the present state. Strange observers which according to such a fluctuation could spontaneously pop into existence have recently been discussed in the context of a "multiverse" picture and there been called "Boltzmann brains", cf. [3] and the references therein. The multiverse picture can be motivated by inflationary scenarios of the early Universe (e.g. Linde's "eternal inflation") and describes the full Universe as being infinitely extended and very inhomogeneous on large scales, but containing many Friedmann subuniverses of the kind that we observe. In such a gigantic Universe, even the tiniest entropic fluctuation would occur somewhere. I shall briefly address the multiverse picture below, but focus in the following on the observable part of the Universe, which is approximately homogeneous and isotropic.

In order to know how special our Universe in fact is, one would like to calculate both the actual entropy of our Universe as well as the maximal possible entropy. The non-gravitational entropy is dominated by the photons of the Cosmic Microwave Background (CMB) radiation; it contributes about $2\times 10^{89}\, k_{\rm B}$ [4]. Linde and Vanchurin have, moreover, given an estimate of an upper limit for the non-gravitational entropy, which would be obtained if all particles were ultra-relativistic: their value is about $10^{90}\, k_{\rm B}$ and thus only about one order of magnitude more than the CMB value [5]. These are very large numbers, but they are much smaller than the gravitational contribution to the entropy. Unfortunately, a general expression for gravitational entropy does not exist. Because of the universal attractivity of gravity, one can only expect that gravitational entropy increases during a gravitational collapse, in contrast to the entropic trend of ordinary matter which prefers a homogeneous state. However, for the most extreme case of gravitational collapse, an entropy formula exists: the Bekenstein–Hawking entropy for black holes. It reads

$$S_{\rm BH} = \frac{k_{\rm B}c^3 A}{4G\hbar} = k_{\rm B}\frac{A}{4l_{\rm P}^2}, \qquad (1)$$

where A is the surface area of the event horizon and $l_{\rm P} = \sqrt{G\hbar/c^3}$ is the Planck length; in the following estimates we shall set Boltzmann's constant $k_{\rm B}$ equal to one.

To see how large the Bekenstein–Hawking entropy can become, let us estimate its value for the Galactic Black Hole – the supermassive black hole in the centre of our Milky Way with a mass $M \approx 3.9 \times 10^6\, M_\odot$. Neclecting its angular momentum, which would anyway decrease the estimated entropy, one gets from (1)

$$S_{\rm GBH} = \pi \left(\frac{R_{\rm S}}{l_{\rm P}}\right)^2 \approx 6.7 \times 10^{90}, \qquad (2)$$

where $R_{\rm S}$ denotes the Schwarzschild radius. This already exceeds by more than one order of magnitude the non-gravitational contribution to the entropy of the observable Universe. According to a recent investigation [4], all supermassive black holes together yield an entropy of $S = 3.1^{+3.0}_{-1.7} \times 10^{104}$.

Roger Penrose has pointed out in [6] that the maximal entropy for the observable Universe would be obtained if all its matter were assembled into one black hole. Taking the most recent observational data, this would yield the entropy (calculated in the Appendix)

$$S_{\rm max} \approx 1.8 \times 10^{121}. \qquad (3)$$

This may not yet be the maximal possible entropy. Our Universe exhibits currently an acceleration which could be caused by a cosmological constant Λ. If this is true, it will expand forever, and the entropy in the far future will be dominated by the entropy of the cosmological event horizon (the "Gibbons–Hawking entropy" [7]). The estimate, presented in the Appendix, yields

$$S_{\rm GH} = \frac{3\pi}{\Lambda l_{\rm P}^2} \approx 2.88 \times 10^{122}, \qquad (4)$$

which is about one order of magnitude higher than (3).

Following the arguments in [6], the "probability" for our Universe can be estimated as

$$\frac{\exp(S)}{\exp(S_{\rm max})} \approx \frac{\exp(3.1 \times 10^{104})}{\exp(2.9 \times 10^{122})} \approx \exp(-2.9 \times 10^{122}). \qquad (5)$$

Our Universe is thus very special indeed. It must have "started" near the Big Bang with an extremely low entropy; the Universe must have been very smooth in the past, with no white holes being present. Penrose has reformulated this observation in his Weyl-tensor hypothesis: the Weyl tensor is zero near the Big Bang (describing its smooth state), but diverges in a Big Crunch (provided the Universe will recollapse). Since the Weyl tensor describes in particular gravitational waves, this hypothesis entails that all gravitational waves must be retarded. This is analogous to the Sommerfeld condition in electrodynamics and the absence of advanced electromagnetic waves [1]. There, the electromagnetic arrow can be traced back to the thermodynamic arrow and the Second law by using the thermodynamic properties of absorbers, but this is not possible here because the absorption cross section of gravitational waves is too small. Still, we shall argue in the last section that all arrows of time can be traced back to one common origin – a universal boundary condition of low initial cosmic entropy.

The Weyl-tensor hypothesis is not yet an explanation, but only a description of the low initial cosmic entropy. Penrose has recently reformulated his hypothesis in the framework of his "Conformal Cyclic Cosmology" (CCC) [8]. A central role is attributed therein to a proposed information loss in black holes and the ensuing nonunitary evolution. His whole picture, however, remains classical as far as gravity is concerned. Here, I would like to adopt instead the point of view that the gravitational field is fundamentally of quantum nature. This is not a logical necessity, but one can put forward physical arguments in favour of quantum gravity as the more fundamental theory [9, 10]. Although there exist non-singular classical solutions, the singularity theorems of classical relativity suggest the abundance of singularities in the classical theory; a more general framework is therefore needed to exorcize them. Moreover, gravity acts universally, so it couples to all other fields of Nature, all of which are so far described by quantum theory. It could then at least be considered unnatural to leave the gravitational field classical; this would become especially awkward in the context of a unified theory of all interactions.

In the next section I shall briefly describe an approach to quantum gravity which is very conservative and which despite its limits should be able to provide insights into the origin of the arrow of time.

3 Quantum Geometrodynamics

A full quantum theory of gravity remains elusive [9]. Can one nevertheless say something reliable about quantum gravity without knowing the exact theory? In [11] I have made the point that this is indeed possible. The situation is analogous to the role of the quantum mechanical Schrödinger equation. Although this equation is not fundamental (it is non-relativistic, it is not field-theoretic), important insights can be drawn from it. For example, in the case of the hydrogen atom, one has to impose boundary conditions for the wave function at the origin $r \to 0$, that is, at the centre of the atom. This is certainly not a region where one would expect non-relativistic quantum mechanics to be exactly valid, but its consequences, in particular the resulting spectrum, are empirically correct to an excellent approximation.

Erwin Schrödinger has found his equation by "guessing" a wave equation from which the Hamilton–Jacobi equation of classical mechanics can be recovered in the limit of small wavelengths, analogously to the limit of geometric optics from wave optics. The same approach can be applied to general relativity. One can start from the Hamilton–Jacobi version of Einstein's equations and "guess" a wave equation from which they can be recovered in the classical limit. The only assumption that is required is the universal validity of quantum theory, that is, its linear structure. It is not yet needed for this step to impose a Hilbert-space structure. Such a structure is employed in quantum mechanics because of the probability interpretation for which one needs a scalar product and its conservation in time (unitarity). The status of this interpretation in quantum gravity remains open, see below.

The result of this approach is quantum geometrodynamics. Its central equation is the Wheeler–DeWitt equation, first discussed by Bryce DeWitt and John Wheeler in the 1960s. In a short notation, it is of the form

$$\hat{H}\Psi = 0, \tag{6}$$

where $\hat{H}$ denotes the full Hamiltonian for both the gravitational field (here described by the three-metric) as well as all non-gravitational fields. For the detailed structure of this equation I refer, for example, to the classic paper by DeWitt [12] or the general review in [9]. Two properties are especially important for our purpose here. First, this equation does not contain any classical time parameter t. The reason is that spacetime as such has disappeared in the same way as particle trajectories have disappeared in quantum mechanics; here, only space (the three-geometry) remains. Second, inspection of $\hat{H}$ exhibits the local hyperbolic structure of the Hamiltonian, that is, the Wheeler–DeWitt equation possesses locally the structure of a Klein–Gordon equation. In the vicinity of Friedmann universes, this hyperbolic structure is not only locally present, but also globally. One can thus define a new time variable which exists only intrinsically and which can be constructed from the three-metric (and non-gravitational fields) itself. It is this absence of external time that could render the probability interpretation and the ensuing Hilbert-space structure obsolete in quantum gravity, for no conservation of probability may be needed.[1]

How, then, can one understand the emergence of an arrow of time from a fundamental equation which is itself timeless? I shall address this issue in the next section.

4 Arrow of Time from Quantum Cosmology

Quantum cosmology is the application of quantum theory to the Universe as a whole. In a first approximation, the Universe is homogeneous and isotropic. The three-metric is then fully characterized by the scale factor, a, of the Universe. Classically, the Universe evolves in time, $a(t)$; the same holds for the matter fields. In quantum cosmology, t has disappeared and all available information is encoded in the wave function $\psi(a, \ldots)$, where the ... denote homogeneous matter degrees of freedom. For the simple two-dimensional configuration space consisting of the scale factor and a minimally coupled scalar field ϕ, the Wheeler–DeWitt equation

[1]The situation is different for an isolated quantum gravitational system such as a black hole; there, the semiclassical time of the rest of the Universe enters the description [13].

reads (with $c = 1$)

$$\hat{H}\Psi = \left(\frac{2\pi G\hbar^2}{3}\frac{\partial^2}{\partial\alpha^2} - \frac{\hbar^2}{2}\frac{\partial^2}{\partial\phi^2} + e^{6\alpha}\left(V(\phi) + \frac{\Lambda}{8\pi G}\right) - 3e^{4\alpha}\frac{k}{8\pi G}\right)\Psi(\alpha,\phi) = 0, \tag{7}$$

with cosmological constant Λ and curvature index $k = \pm 1, 0$. The variable $\alpha = \ln a$ has been introduced for convenience.

In order to discuss thermodynamical issues, additional degrees of freedom must be added. One option is to consider small inhomogeneities in the vicinity of homogeneity. This can be achieved, for example, through a multipole expansion on the three-sphere (assuming the Universe is closed) [14, 15]. Schematically, the Wheeler–DeWitt equation (6) then assumes the form

$$\hat{H}\,\Psi = \left(\frac{2\pi G\hbar^2}{3}\frac{\partial^2}{\partial\alpha^2} + \sum_i \left[-\frac{\hbar^2}{2}\frac{\partial^2}{\partial x_i^2} + \underbrace{V_i(\alpha, x_i)}_{\to 0 \text{ for } \alpha\to-\infty}\right]\right)\Psi = 0, \tag{8}$$

where the $\{x_i\}$ denote the scalar field as well as the inhomogeneous degrees of freedom; $V_i(\alpha, x_i)$ are the corresponding potentials. One recognizes immediately that this Wheeler–DeWitt equation is hyperbolic with respect to the intrinsic time α. Initial conditions are thus most naturally formulated with respect to constant α.

The important observation is now that the potential in (8) is *asymmetric* with respect to α; if written out, it contains explicit factors of $e^{6\alpha}$, etc., and vanishes in the limit $\alpha \to -\infty$. In contrast to almost all the other fundamental equations in physics, it thereby distinguishes a direction in (intrinsic) time. One could thus envisage a solution to the Wheeler–DeWitt equation, that near the Big Bang would be an approximate product state between all degrees of freedom [1],

$$\Psi \overset{\alpha\to-\infty}{\longrightarrow} \psi_0(\alpha)\prod_i \psi_i(x_i). \tag{9}$$

Introducing the entropy of the Universe as an entanglement entropy, in which irrelevant, that is, unobservable or unobserved degrees of freedom (such as small gravitational waves described by some of the x_i) are integrated out, the state (9), which is a product state, would yield a vanishing entropy. For increasing α, this solution would evolve into a superposition of α and the inhomogeneous modes (as well as between the inhomogeneous modes). Integrating out all or part of the x_i would then yield a non-vanishing and increasing entropy with respect to α. Increasing entanglement would then cause increasing decoherence for the relevant degrees of freedom [15–17]. Decoherence – the unavoidable emergence of classical properties through interaction with irrelevant degrees of freedom – is perhaps the most fundamental irreversible process and thus stands behind all arrows of time

[1, 17]. Because of the asymmetry of (8) with respect to α, substituting $\alpha \to -\alpha$ would *not* yield a solution of the Wheeler–DeWitt equation. If a solution of the form (8) were the one describing our Universe, we could understand from it the irreversible appearance of our world. A full understanding of quantum gravity would perhaps single out a unique solution for the Wheeler–DeWitt equation, a possibility already envisioned by DeWitt [12].

There are indications that the above quantum state would evolve into a symmetric state where all perturbations are in an (at least approximate) de Sitter-invariant vacuum state. Such a vacuum state is a good candidate for the early Universe [18]. The state for each perturbation mode would describe a superposition of inhomogeneous states, that is, a non-classical state. However, the mechanism of decoherence also comes into play here, generating a classical behaviour for the modes which may then serve as the seeds for the origin of structure in the Universe [19].

We have not yet discussed the connection between the intrinsic time α and the "observed time" t which should be at our disposal at least in an appropriate semiclassical limit. This can be achieved through a Born–Oppenheimer type of approximation scheme, cf. [9] and the references therein. Some degrees of freedom such as the scale factor may serve as the semiclassical variables from which a semiclassical time t can be defined in appropriate situations. This will be the time variable which controls the dynamics in this approximation and which enters an effective and approximate Schrödinger equation for the non-gravitational quantum variables. The arrow of time aligned along increasing scale factor a thus trivially extends to the semiclassical time t – as long as the semiclassical approximation is valid.

The above ideas may, with slight elaborations, also apply to the idea of the multiverse, that is, to a Universe with many approximately homogeneous sub-universes, cf. [5] and the references therein. Quantum entanglement is not limited to sub-horizon scales and may thus be effective also in the full multiverse. Decoherence should then distinguish the same arrow of time everywhere in the multiverse. Applying also here the idea that quantum fluctuations, after their effective classicality due to decoherence, become the seeds for galaxy formation, Linde and Vanchurin estimate in [5] the number of realizations of the emergent classical fluctuations in the gigantic multiverse. This number would also correspond to the number of branches of the universal wave function in the Everett interpretation when applied to our Hubble domain. After decoherence, each realization can serve as a classical initial condition for the subsequent evolution of the Universe. They find for the total number of distinguishable locally "Friedmann universes" the estimate

$$e^{S_{\rm pert}} \lesssim e^{e^{3N}} , \tag{10}$$

where $S_{\rm pert}$ is the total entropy of the perturbations, see also [19, 20], and N is the number of e-folds of slow-roll (post-eternal) inflation. In the simplest models of chaotic inflation, one thereby gets the incredibly high number [5]

$$10^{10^{10^7}} .$$

(A much lower number – two instead of three exponentials – is obtained in the case of a positive cosmological constant.) Adopting, in addition, the landscape picture of string theory, this estimate would correspond to the case of one vacuum. Taking all the vacua into account, the number will be even higher. The issue of the Wheeler–DeWitt equation on a configuration space mimicking the landscape picture and the question of a low-entropy initial condition was discussed, for example, in [21].

The idea of quantum cosmology is that the whole Universe at all scales is described by quantum theory. Thus, a priori, quantum effects are not restricted to the Planck scale. In the case of a classically recollapsing quantum universe, for example, one can predict the occurrence of quantum effects near the classical turning point [22], see also [1, 23] for a detailed explanation. Because the arrow of time in the above scenario is correlated with increasing scale factor a, the quantum universe would in this case consist of many branches in which the arrow of time always points in the direction of increasing a. These branches would be decohered components of the universal wave functions and would thus become independent of each other for most of their existence, but they would interfere destructively at the classical turning point in order to fulfill the boundary condition $\Psi \to 0$ for $a \to \infty$, which is necessary for a model in which the classical trajectories recollapse. Consequently, no classical observers would be able to survive a transition through the turning point, and time as well as the classical evolution would come to an end there.

This is an impressive example for a quantum effect at large scales. Other examples can be found in models that are of interest because they may describe a dynamical dark energy in our Universe. Examples are models which classically exhibit a big-rip or a big-brake singularity; in the first case, the Universe can become infinitely large in a finite time, while in the second case it comes to an abrupt halt in the future. In both cases, this corresponds to a singular region. One can now study solutions of the corresponding Wheeler–DeWitt equations and finds that the singularities will be avoided, cf. [23] and the references therein: the semiclassical approximation breaks down when approaching the region of the classical singularity, and for the big brake the wave function even becomes zero there. What are the consequences of this scenario for the arrow of time?

Since the semclassical time comes to an end, so does the arrow. The Universe enters a genuine quantum era which no classical observers (and others are not known) could survive. This is analogous to the above discussed turning point. The world then becomes truly timeless.

One might wonder what happens in the case of models which classically describe bouncing cosmologies [24]: the Universe would then undergo many, perhaps infinite, cycles of expansion and recollapse. What would happen with the entropy in these cases? If the entropy were indeed correlated with the scale factor, as the scenario discussed above suggests, the arrow of time would not continue through a turning point. The bouncing models would thus make no sense in quantum cosmology; one would only have branches of the wave function in which the arrow would point from small to large universe and where time would end when a classical turning point is approached.

We have restricted the discussion to quantum geometrodynamics. At least for scales above the Planck length, which includes the above discussed quantum scenarios for a big universe in the future, this should provide a reliable framework. Modifications are, however, expected when approaching the Planck regime, that is, the region of the Big Bang. Such modifications have been addressed in string theory and loop quantum gravity [9]. In the case of loop quantum cosmology, the Wheeler–DeWitt equation is replaced at the most fundamental level by a difference (instead of differential) equation. For a big universe, the differences to quantum geometrodynamics are negligible; this concerns, for example, the examples of the big rip and the big brake. Near the big-bang singularity, however, the situation is different. The emerging scenario is discussed at length in another contribution to this volume [25]. Also there, the author suggests "the possibility of deriving a beginning within a beginningless theory". Thus, although the approaches may be different, the common fundamental challenge is to understand the observed time and its arrow from a scenario of the world which is fundamentally timeless.

Acknowledgements I thank Max Dörner and Tobias Guggenmoser for a careful reading of this manuscript.

Appendix: Some Numerical Estimates

Here we recapitulate the numerical estimates about the maximal possible entropy in the Universe, presented by Penrose in [6], in the light of recent cosmological data [26]. Since our Universe is spatially flat to a high degree of accuracy, the mass of the matter (both visible and dark) in our present Hubble volume is given by

$$M_{\mathrm{U}} = \frac{4\pi\rho_{\mathrm{m}}c^3}{3H_0^3}, \tag{11}$$

where ρ_{m} is the matter density, and $H_0 \approx 70.5$ km/s Mpc $\approx 2.27 \times 10^{-18}\ \mathrm{s}^{-1}$ is the Hubble constant. Introducing the critical density

$$\rho_{\mathrm{c}} = \frac{3H_0^2}{8\pi G}, \tag{12}$$

we can use the density parameter $\Omega_{\mathrm{m}} = \rho_{\mathrm{m}}/\rho_{\mathrm{c}} \approx 0.274$ and write

$$M_{\mathrm{U}} = \frac{c^3\Omega_{\mathrm{m}}}{2GH_0}. \tag{13}$$

In order to estimate the maximal entropy, we shall assume that the Universe up to the Hubble scale consists of one black hole with mass M_{U}. Since our present Universe is dominated by dark energy, which for our purpose here can be approximated by a cosmological constant Λ, we have to take into account that the metric outside this hole is, in fact, the Schwarzschild–de Sitter metric, see, for example, [27]. Numerically we have $\Lambda \approx 1.25 \times 10^{-56}\,\mathrm{cm}^{-2}$ and $\Omega_\Lambda = \Lambda c^2/3H_0^2 \approx 0.726$ [26]. In the Schwarzschild–de Sitter metric, the black hole has a maximal mass given by

$$M_{\mathrm{N}} = \frac{c^3}{3\sqrt{3\Omega_\Lambda}GH_0} \approx 4 \times 10^{55}\ \mathrm{g}, \tag{14}$$

which corresponds to the case of the Nariai metric (therefore the index N). We thus have to check whether M_{U} is greater or smaller than M_{N}; only in the latter case can the Universe accommodate one single black hole. A short calculation shows

$$\frac{M_{\mathrm{U}}}{M_{\mathrm{N}}} = \frac{3\sqrt{3\Omega_\Lambda}\Omega_{\mathrm{m}}}{2} \approx 0.61, \tag{15}$$

so all of M_{U} can indeed be assembled into one black hole.

We now assume that the maximal entropy is given by a Schwarzschild black hole with mass M_{U} (a non-vanishing angular momentum would give a smaller entropy). Thus,

$$S_{\max} = \pi \left(\frac{R_{\mathrm{h}}}{l_{\mathrm{P}}}\right)^2, \tag{16}$$

where R_{h} denotes the radius of the black-hole event horizon (as opposed to the cosmological horizon R_{c}). In the Schwarzschild–de Sitter metric we have

$$R_{\mathrm{h}} = \frac{3GM_{\mathrm{U}}\ell\xi}{c^2}\left(1 - \sqrt{1 - \frac{1}{\ell\xi^3}}\right) = \frac{\xi}{\sqrt{\Lambda}}\left(1 - \sqrt{1 - \frac{1}{\ell\xi^3}}\right), \tag{17}$$

where $\ell^{-1} = M_{\mathrm{U}}/M_{\mathrm{N}}$ and $\xi = \cos(\frac{1}{3}\cos^{-1}[\ell^{-1}]) \approx 0.95$. With the above numbers we have

$$R_{\mathrm{h}} \approx 2.13\frac{GM_{\mathrm{U}}}{c^2} \approx 0.38 \times 10^{28}\ \mathrm{cm} \tag{18}$$

and therefore

$$S_{\max} \approx 1.8 \times 10^{121}. \tag{19}$$

This is the number that should replace the estimate 10^{123} in [6] if one makes use of the data presented in [26].

Expressed in grams, the mass (11) is $M_{\mathrm{U}} \approx 2.4 \times 10^{55}$ g and would therefore correspond to about 1.5×10^{79} baryons if all mass were in baryons (the actual number is smaller because of the non-baryonic dark matter). In the case of 10^{80} baryons, as used in [8], one would find $\ell^{-1} > 1$, that is, the corresponding mass would exceed the Nariai mass (14) and it would thus not be possible to assemble this mass into a single black hole.

In the case of a non-vanishing cosmological constant there occurs also the gravitational entropy S_Λ associated with the cosmological event horizon R_c [7], where

$$R_c = \frac{3GM_U\ell\xi}{c^2}\left(1+\sqrt{1-\frac{1}{\ell\xi^3}}\right) \approx 1.29\times 10^{28}\ \text{cm}. \tag{20}$$

It reads

$$S_\Lambda = \pi\left(\frac{R_c}{l_P}\right)^2 \approx 1.99\times 10^{122}. \tag{21}$$

One should thus in principle consider the sum of S_Λ and the entropy associated with all matter being trapped in a single black hole. However, the maximal entropy is reached for asymptotic times $t \to \infty$ when the matter content becomes irrelevant (because it will be diluted and no black hole with mass M_U will be formed); the radius of the event horizon then approaches

$$R_c = \sqrt{\frac{3}{\Lambda}} \approx 1.55\times 10^{28}\ \text{cm}. \tag{22}$$

The entropy associated with this event horizon then approaches the "Gibbons–Hawking" entropy [7]

$$S_{GH} = \frac{3\pi}{\Lambda l_P^2} \approx 2.88\times 10^{122}, \tag{23}$$

which is about 16 times the black-hole maximal entropy (19). The numerical value in (23) is also presented in [4].

Taking the case of the Nariai mass (14), one would have the total entropy $S_N + S_\Lambda = 2S_{GH}/3$, which would give further support to consider (23) as the maximal possible entropy of the observable Universe, as suggested by current observational data.

References

1. H.D. Zeh, *The Physical Basis of the Direction of Time*, 5th edn. (Springer, Berlin, 2007)
2. H.D. Zeh, Open questions regarding the arrow of time (2012). Contribution to this volume
3. A. De Simone, A.H. Guth, A. Linde, M. Noorbala, M.P. Salem, A. Vilenkin, Boltzmann brains and the scale-factor cutoff measure of the multiverse Phys. Rev. D **82**, 063520 (2010) [arXiv:0808.3778v1 [hep-th]]
4. C.A. Egan, C.H. Lineweaver, A larger estimate of the entropy of the universe Astrophys. J. **710**, 1825–1834 (2010) [arXiv:0909.3983v1 [astro-ph.CO]]

5. A. Linde, V. Vanchurin, How many universes are in the multiverse? Phys. Rev. D **81**, 083525 (2010) [arXiv:0910.1589v1 [hep-th]]
6. R. Penrose, Time-asymmetry and quantum gravity. In *Quantum Gravity*, vol. 2, ed. by C.J. Isham, R. Penrose, D.W. Sciama (Clarendon Press, Oxford, 1981), pp. 242–272
7. G.W. Gibbons, S.W. Hawking, Cosmological event horizons, thermodynamics, and particle creation. Phys. Rev. D **15**, 2738–2751 (1977)
8. R. Penrose, Black holes, quantum theory and cosmology. J. Phys. Conf. Ser. **174**, 012001 (2009)
9. C. Kiefer, *Quantum Gravity*, 2nd edn (Oxford University Press, Oxford, 2007)
10. M. Albers, C. Kiefer, M. Reginatto, Measurement analysis and quantum gravity. Phys. Rev. D **78**, 064051 (2008)
11. C. Kiefer, Quantum geometrodynamics: whence, whither? Gen. Relativ. Gravit. **41**, 877–901 (2009); C. Kiefer, Does time exist in quantum gravity? (2009) [arXiv:0909.3767v1 [gr-qc]]
12. B.S. DeWitt, Quantum theory of Gravity. I. The canonical theory. Phys. Rev. **160**, 1113–1148 (1967)
13. C. Kiefer, J. Marto, and P.V. Moniz (2009): Indefinite oscillators and black-hole evaporation. *Ann. Phys. (Berlin)* **18**, 722–735.
14. J.J. Halliwell, S.W. Hawking, Origin of structure in the universe. Phys. Rev. D **31**, 1777–1791 (1985)
15. C. Kiefer, Continuous measurement of minisuperspace variables by higher multipoles. Class. Quantum Grav. **4**, 1369–1382 (1987)
16. H.D. Zeh, Emergence of classical time from a universal wave function. Phys. Lett. A **116**, 9–12 (1986)
17. E. Joos, H.D. Zeh, C. Kiefer, D. Giulini, J. Kupsch, I.-O. Stamatescu, *Decoherence and the Appearance of a Classical World in Quantum Theory*, 2nd edn (Springer, Berlin, 2003)
18. A.A. Starobinsky, Spectrum of relict gravitational radiation and the early state of the universe. JETP Lett. **30**, 682–685 (1979)
19. C. Kiefer, I. Lohmar, D. Polarski, A.A. Starobinsky, Pointer states for primordial fluctuations in inflationary cosmology. Class. Quantum Grav. **24**, 1699–1718 (2007); C. Kiefer, D. Polarski, Why do cosmological perturbations look classical to us? Adv. Sci. Lett. **2**, 164–173 (2009) [arXiv:0810.0087v2 [astro-ph]]
20. C. Kiefer, Entropy of gravitational waves and primordial fluctuations. In *Cosmology and Particle Physics*, ed. by J. Garcia-Bellido, R. Durrer, M. Shaposhnikov (American Institute of Physics, New York, 2001), pp. 499–504
21. R. Holman, L. Mersini-Houghton, Why the universe started from a low entropy state. Phys. Rev. D **74**, 123510 (2006)
22. C. Kiefer, H.D. Zeh, Arrow of time in a recollapsing quantum universe. Phys. Rev. D **51**, 4145–4153 (1995)
23. C. Kiefer, B. Sandhöfer, Quantum cosmology. In *Beyond the Big Bang*, ed. by R. Vaas (Springer, Berlin, 2012), [arXiv:0804.0672v2 [gr-qc]]
24. M. Novello, S.E.P. Bergliaffa, Bouncing cosmologies. Phys. Rep. **463**, 127–213 (2008)
25. M. Bojowald, A momentous arrow of time (2012). Contribution to this volume
26. G. Hinshaw et al., Five-year Wilkinson microwave anisotropy probe (WMAP) observations: Data processing, sky maps, and basic results. Astrophys. J. Suppl. **180**, 225–245 (2009)
27. K.H. Geyer, Geometrie der Raum-Zeit der Maßbestimmung von Kottler, Weyl und Trefftz. Astron. Nachr. **301**, 135–149 (1980)

Open Questions Regarding the Arrow of Time

H. Dieter Zeh

Abstract Conceptual problems regarding the arrow of time in classical physics, quantum physics, cosmology, and quantum gravity are discussed. Particular attention is paid to the retardation of various kinds of correlations, the dynamical rôle of the quantum indeterminism, and to different concepts of timelessness.

1 Laws and Facts

The Second law of thermodynamics is usually regarded as the major physical manifestation of the arrow of time, from which many other consequences can be derived. I have discussed the relations between these different forms of the arrow in detail elsewhere [1], so I will occasionally refer to this source in the following by TD ("Time Direction") for short. This article is meant to review some open conceptual problems, which are often insufficiently realized, or of actual interest for other reasons.

In statistical thermodynamics, the Second law is derived from the assumption that a closed system must evolve towards a more probable state. In this context, entropy is defined as a measure of probability. This explanation is incomplete for various reasons. First, the concept of evolution already presumes a direction in time. To regard it as a direction *of* time would even apply this asymmetry to the very definition of time. This would go beyond a purely mechanistic concept of time, which is defined in accordance with time-symmetric laws of motion. Newton's absolute "flow of time" is a similar metaphor; its direction would be physically meaningful only if one assumed asymmetric laws. For example, Newton regarded friction as representing a fundamental force that would slow down all motion.

H.D. Zeh (✉)
Universität Heidelberg, Gaiberger Str. 38, D-69151 Waldhilsbach, Germany
e-mail: zeh@uni-heidelberg.de

L. Mersini-Houghton and R. Vaas (eds.), *The Arrows of Time*, Fundamental Theories of Physics 172, DOI 10.1007/978-3-642-23259-6_11,

So in his opinion God had to intervene once in a while to set things in motion again. Without such an external, metaphysical, or at least law-like fundamental direction in or of time, one can only speak of an asymmetry of the facts (which may well be compatible with symmetric laws).

Second, the concept of probabilities requires a measure that is usually defined with respect to ensembles of *possible* states. Since in classical physics every system is assumed at any time to be in one definite microscopic state, the latter must be "coarse-grained" in order to define a macroscopic or "thermodynamic state", that is, an ensemble of microscopic states which may thus define a non-trivial probability measure. Various kinds of coarse-graining (omissions of actual or possible information) have been discussed in the rich literature on this subject, or were invented in the context of a new theory. The justification of such ensembles by a macroscopic (that is, incompletely defined) preparation procedure would refer to the time-directed concept of preparations as a *deus ex machina* (similar to Newton's divine interventions). Arguments based on incomplete observability or controllability of certain degrees of freedom may also presume external time-asymmetric observers.

The mechanistic concept of time is usually postulated together with the deterministic dynamical laws that are assumed on empirical grounds to control the facts. Eugene P. Wigner called the distinction between laws and initial conditions (initial facts) Newton's greatest discovery. In a deterministic theory, initial conditions could as well be replaced by final ones, or by conditions at any intermediate time. This mechanistic concept requires only that time can be represented by the real numbers (without any preference for their sign) – thus defining a linear order of physical states or global "Nows". Deterministically, the size of an ensemble (the number of microscopic states, or an appropriate measure if this number is infinite) does not change in time, while its coarse-grained size would, in general. This is why Ludwig Boltzmann's statistical measure H, which up to a factor may be assumed to represent "negentropy" for a diluted gas, may change in time (see Sect. 2). It will decrease even under deterministic equations of motion *in the direction of calculation* – provided the fixed concept of coarse graining was used to define the input ensemble. Further conditions studied in ergodic theory are necessary to exclude exceptional cases that are usually of measure zero. There is no a priori reason to calculate only in the conventional "forward" direction of time, but this is *empirically* the only direction in which statistical arguments lead to correct results – thus indicating a strong asymmetry of the facts. For applications to cosmology let me emphasize that the conservation of exact (not coarse-grained) ensemble entropy under deterministic equations of motion would include the situation of a deterministically inflating universe, which has often erroneously be claimed to *cause* a low entropy condition.

Although irreversible phenomena are mostly observed locally, the thermodynamical arrow of time seems to possess a common global direction. Its origin has, therefore, usually been discussed in a cosmological context. For example, one may assume a special cosmic initial condition at the big bang. Boltzmann, who assumed the universe to be eternal, argued instead that a giant chance fluctuation must have occurred in the distant past in order to form a low entropy state. A physical "future" would then be characterized by any time direction away from such a low entropy state. Boltzmann's proposal seems to imply, though, that it would be far more probable to assume that the *present* state of the universe – including all memories and conscious brains – had just formed in a chance fluctuation, since this state would possess very low, but nonetheless much higher entropy than the otherwise required state in the distant past (see, however, Sect. 2). This idea has recently been discussed under the name "Boltzmann brains", mainly in some as yet speculative cosmologies. If, on the other hand, the low entropy is related to a global special condition at high densities, the thermodynamical time arrow would have to change direction in an oscillating universe, while simultaneous opposite arrows in causally connected parts of the universe seem to be excluded for dynamical reasons [2, 3].

The arguments based on deterministic dynamics do not directly apply to stochastic dynamics. However, a stochastic law by itself does not necessarily characterize a direction in time. If all states at some time t_1 had two possible successors at a later time t_2, say, and if this law held on all states, then each successor must on average also have two dynamically possible predecessors at time t_1. Therefore, such a stochastic law defines a time-asymmetric indeterminism only when applied to a genuine subset of possible *initial* states, while not restricting the set of final states. This would be just another way of applying the "double standard" that has been duly criticized by Price [4]. The asymmetry is not a consequence of the stochastic law itself (see Sect. 3.4 of TD and the concept of "forks of indeterminism" mentioned therein). On the other hand, even deterministic laws may be asymmetric, but this would not by itself offer a way to explain the increase of entropy. Examples are the Lorentz force of an external magnetic field or *CP*-violation. In these and similar cases, formal time-reversal symmetry violation is compensated for by another symmetry violation, which may be either physical, such as a *CP* transformation, or just formal, such as complex conjugation in the Schrödinger equation.

Our world is known to obey quantum theory, which is characterized by an indeterminism occurring in measurements and other "quantum events". There is absolutely no consensus among physicists about the interpretation and even the precise dynamical rôle of this "irreversible coming into being" of the observed facts, such as the click of a counter. Has it to be regarded as a specific part of the dynamical laws (as assumed in the form of von Neumann's "first intervention" or more explicitly in collapse theories), as representing events that (according to Wolfgang Pauli) occur outside the laws of nature, as a "normal" increase of information (as claimed in the Copenhagen interpretation), as determined by hidden variables that are not counted in conventional ensemble entropy (as in David Bohm's theory), or as the consequence of inderministically splitting quantum observers (as in Hugh Everett's interpretation)? Some quantum cosmologists refer to initial uncertainty relations or "quantum fluctuations" in order to justify the stochastic evolution of their quantum universe, although a global quantum state is never required to be "uncertain" (only classical variables had to be assumed to be uncertain if they were used).

In the pragmatic Copenhagen interpretation, this problem is essentially circumvented by denying any microscopic reality, while other above-mentioned proposals suggest novel laws or concepts, which may or may not be confirmed or ruled out in principle. Although these various interpretations must have drastic consequences for the resulting model of the universe, they play surprisingly almost no rôle in actual cosmology. For example, the thermodynamical arrow might be the consequence of a time-asymmetric collapse mechanism if this were part of the laws. In the Copenhagen interpretation, there simply "is no quantum world" – hence no complete cosmological model. Most cosmologies are therefore based on classical concepts, just allowing for some "quantum corrections", while indeterministic master equations are often derived from unitary equations of motion by using certain "approximations" in analogy to classical statistical physics. Such equations may then even *appear to explain* stochastic and irreversible quantum events, although they are instead implicitly using them.

Much philosophical debate has also been invested into the pseudo-distinction between a block universe and an evolving universe (a world of being versus a world of becoming). However, these apparently different pictures describe only different representations of the same thing. One should realize that a block universe picture is by no means restricted to a physical context. Historians have always been applying it to the past, although they never had doubts that Cesar crossed the Rubicon according to his free will. We can similarly use space-time diagrams to represent *actual* motions or *potential* histories (individual members of an ensemble of possible histories) even in the case of an indeterministic law. Moreover, a block universe picture has nothing specifically to do with relativity (except that it is just convenient in the absence of a concept of absolute simultaneity).

2 The Arrow in Classical Physics

It is essential to keep in mind that time-symmetric laws are perfectly compatible with asymmetric solutions. Almost all solutions of the fundamental equations of motion are time-asymmetric, while reasonably defined quasi-recurrence times for isolated systems would exceed the age of the universe by enormous factors. The symmetry of the laws of motion requires only that for every asymmetric solution that is realized in nature there must mathematically – not necessarily physically – exist precisely another, time-reversed one. In reality, though, very few systems can be considered as being isolated [5]. This means that the reversed solution would require an exact time reversal of its complete environment – an argument that must then be extended to the whole causally connected region of our universe. An extremely small "perturbation" (change of the state at some time) would with overwhelming probability turn a deterministic solution with decreasing entropy into one with increasing entropy (in both directions of time) [2].

Remarkable is only that there are whole classes of asymmetric solutions that are found in abundance, while members of the reversed class are rarely or never observed. As an example, consider the contrast between retarded and advanced Maxwell fields for a given type of source. This asymmetry may be understood as a consequence of the presence of absorbers (including the early radiation era of our universe). Absorbers are based on the thermodynamical arrow of time, since they describe the transition to thermal equilibrium between radiation and matter. So they produce "retarded shadows", which, when forming a complete spatial boundary, give rise to local initial conditions of no incoming radiation at frequencies above the thermal spectrum (see Chap. 2 of TD). But why do all physical absorbers absorb in one and the same direction of time only?

The precise microscopic states of systems consisting of many interacting constituents can hardly ever be *known* even in a classical world. So it is common practice to use an incomplete description for them (a generalized coarse-graining). For example, a gas may be described by the mean phase space distribution $\rho_\mu(\boldsymbol{p}, \boldsymbol{q}, t)$ of its molecules. Its evolution in the forward direction of time is then successfully described by Boltzmann's stochastic collision equation. This asymmetric success must be a consequence of properties of the thereby neglected *correlations* between molecules, since the increase of Boltzmann's entropy S_{B},

$$S_B := -Nk_B\overline{\ln \rho_\mu} = -Nk_B \int \rho_\mu \ln \rho_\mu d^3pd^3q, \tag{1}$$

where k_{B} is Boltzmann's constant and N the particle number, can be deterministically understood as a dynamical transformation of information represented by the μ-space distribution into information about correlations. Both kinds of information are described by the $6N$-dimensional Γ-space distribution ρ_Γ, whereby the analogously defined ensemble entropy S_Γ does *not* change under deterministic dynamics. While dynamical models readily confirm that correlations produced in a scattering process remain irrelevant for ρ_μ for an extremely long time, one has to assume asymmetrically that only "retarded correlations", required to reproduce the past, are relevant for the single-particle distribution. This absence of advanced correlations is even "probable", while the low-entropy initial condition that leads to retarded correlations – such as a special initial μ-space distribution ρ_μ is *not*. Explaining this asymmetry by referring to "causality" would beg the question.

There are many appropriate ways to distinguish between macroscopic and microscopic ("irrelevant") degrees of freedom. They can all be formally described by some idempotent "Zwanzig" operator P that acts on the Γ-space distributions $\rho = \rho_\Gamma$ (see Sect. 3.2 of TD),

$$\rho = P_{rel}\,\rho + P_{irrel}\,\rho, \quad \text{with} \quad P_{rel}^2 = P_{rel} \quad \text{and} \quad P_{irrel} = 1 - P_{rel}, \tag{2}$$

where the macroscopically relevant part, $\rho_{\text{rel}} = P_{\text{rel}}\rho$, defines a generalized "coarse-grained" distribution. Macroscopic properties are characterized by a certain robustness or controllability, which may vary with the physical situation. For example, correlations between molecules or ions are stable and relevant in solid bodies, while the corresponding lattice vibrations can then mostly be treated thermally. Although the exact dynamics requires a coupling between ρ_{rel} and ρ_{irrel}, there often exists a probabilistic effective "master equation" for ρ_{rel} that reflects the dynamical future irrelevance of ρ_{irrel} for the dynamics of ρ_{rel}, as exemplified by Boltzmann's collision equation, where ρ_{rel} can be defined in terms of ρ_μ.

The physically appropriate relevance concept used to define ρ_{rel} may thus change in time. In such cases, the usual ignorance of microscopic degrees of freedom can be deterministically transformed into "lacking information" about arising macroscopic ("relevant") ones – such as the positions of droplets formed during a condensation process. This happens, in particular, in symmetry-breaking phase transitions, or in measurements of microscopic variables (but these processes assume a completely new form in quantum theory). Strictly speaking, only the complete ensemble entropy, measured by the mean logarithm of ρ_Γ itself (without any coarse-graining), is conserved under deterministic equations of motion. Physical entropy is usually defined *not* to include that part which represents lacking information about macroscopic variables, but rather as a *function* of them. However, the transformation of physical entropy into entropy of lacking information about variables that are usually assumed to be "physically given" cannot be used in a cyclic process to construct a perpetuum mobile of the second kind [6, 7]. Although the formal entropy of lacking information is in general thermodynamically negligible, it may become essential for fundamental considerations – such as those involving Maxwell's demon. While the precise definition of entropy (its specific relevance concept or Zwanzig projection) is in principle a matter of convenience, the initial cosmic low-entropy condition that would "cause" an arrow of time must represent a specific property of the universe, and its precise nature should therefore be revealed.

The robustness of macroscopic properties together with the retardation of all correlations between them means that there are many redundant macroscopic "documents" (including fossils and personal memories) about the macroscopic past. The latter is therefore said to be "overdetermined" by the macroscopic present or future [8]. It appears fixed because it could not have been different if just one (or a few) documents were found to be different. Precisely this *consistency of the documents* makes them trustworthy and distinguishes them from mere chance fluctuations with the same low value of physical entropy. Julian Barbour has called states that contain consistent documents (regardless of their causal origin) "time capsules" [9]. Since conventional concepts of physical entropy are local (based on an entropy *density*), they cannot distinguish between consistent and inconsistent documents. An evolved ("historical") state has much lower statistical probability than indicated by its physical entropy, and this fact may rule out Boltzmann brains for being "statistically unreasonable" (see Sect. 3.5 of TD).

In most cosmological models, the low-entropy initial condition is represented by a "simple" state of high symmetry – very different from a later state of still low but larger entropy that describes complexity and dynamical order as it exists in organisms, for example. While an exactly symmetric state could not evolve into an asymmetric one by means of symmetric and deterministic laws, a state consisting of classical particles cannot be *exactly* (microscopically) homogenous: the information capacity of a single continuous variable is infinite, and any exact value of a spatial variable would violate homogeneity. Nonetheless, a Laplacean universe that is symmetric after appropriate coarse-graining may determine all later arising complexity.

While, in a laboratory situation, thermal equilibrium normally requires macroscopically homogeneous ensembles, such states are still extremely improbable (and hence unstable) in self-gravitating systems. Gravitating stars and galaxies, for example, possess negative heat capacity: they become hotter and denser when losing energy (see Sect. 5.1 of TD). Classically, this negative heat capacity would even be unbounded. Therefore, the initial homogeneity of the universe is a major candidate for the specific low entropy condition that characterizes this universe. Roger Penrose has formulated this condition in general relativity by postulating a vanishing Weyl tensor on all past singularities. This source-free part of the spacetime curvature tensor can be interpreted as representing gravitational radiation. The Weyl condition would thus mean that all gravitational radiation must be retarded (possess sources in its cosmic past that begins at the singularity). An analogous condition had been proposed for electromagnetic radiation by Planck in a debate with Boltzmann, and later by Ritz in a debate with Einstein. However, because of the weak coupling of gravity to matter, the Weyl tensor condition cannot similarly be explained by the thermodynamic properties of absorbing matter. It may then itself establish the causal nature of the universe, that is, be responsible for the absence of future-relevant early correlations.

3 The Arrow in Quantum Theory and Quantum Cosmology

Although the quantum formalism of irreversible processes is formally quite analogous to its classical counterpart (see Sect. 4.1 of TD), there are at least three genuine quantum aspects that are important for the arrow of time: (1) the superposition principle, (2) a quantum indeterminism of controversial origin – often described by a collapse of the wave function, and (3) quantum nonlocality – a specific consequence of (1).

The superposition principle allows *exactly* symmetric elementary states for all kinds of symmetries. Such symmetric states may then form candidates for an entirely unspecific initial pure state. Although they cannot unitarily evolve into asymmetric states by means of a symmetric Hamiltonian, they could do so by means of an appropriate indeterministic collapse of the wave function that does not obey the principle of sufficient reason. While such a collapse has always to be used *in practice* in order to describe measurements or phase transitions in terms of quantum states, a non-unitary modification of the Schrödinger equation that would satisfactorily describe it in a general way has never been experimentally confirmed. Therefore, Everett's "branching" of the quantum universe (including all observers) into different autonomous components describing quasi-classical "worlds" must be taken seriously as forming an alternative – whatever it means. This branching is objectively specified by an in practice irreversible decoherence process that is described by the Schrödinger equation.

While the collapse would define a time-asymmetric law, the time arrow of decoherence (formation of retarded entanglement) must again arise as a consequence of an initial condition – now for the global wave function. A universally valid Schrödinger equation would in principle also admit the anti-causal process of recoherence, but this is very rare under an appropriate initial condition. Although any initial symmetry of the global state must be conserved under a symmetric Hamiltonian, a non-entangled ("simple") symmetric state can evolve into a symmetric superposition of many asymmetric Everett branches (independent "worlds" possessing a complex structure). This subtlety is neglected in many quantum cosmological models – in particular when other formally arising Everett branches are simply disregarded for being "meaningless". Unitarily calculating backwards in time, however, would require knowledge of *all* Everett branches or collapse components (including the unobserved ones) *and their phase relations* as an input. While the macroscopic past ("history") is overdetermined by the present even in an individual branch, the microscopic past is underdetermined even if all present branches (in conventional language "possibilities" that *could* have occurred) were known independently of one another.

A stochastic collapse by itself (that is, when neglecting the accompanying decoherence processes) would *reduce* nonlocal entanglement, since it is usually defined to select components that factorize in the relevant subsystems (see Sects. 4.6 and 6.1 of TD). This consequence applies as well to the transition into an *individual* Everett world that is experienced by local (themselves branching) observers which are in definite states. Such a dynamical reduction of entanglement is required, in particular, in order to obtain definite outcomes in measurement-like processes, or to allow the preparation of pure initial states in the laboratory or during a process of self-organization.

This indeterministic transition into less entangled states must reduce any physical entropy measure that is defined by means of a Zwanzig projection of locality (as required if entropy is to be an extensive quantity). It is here important to recognize the difference between *classical* microscopic states, which are local by definition (that is, they are defined by the states of all their local subsystems), and generically nonlocal quantum states. Therefore, the physical (local) entropy of a completely defined ("real") classical state is minimal (minus infinity), while that of a pure quantum state is not only non-negative, but in general also much greater than zero (non-trivial). The permanent creation of uncontrollable quantum entanglement by decoherence must dominate the creation of physical entropy, which would for large times lead to those apparent local ensembles (improper mixtures) that represent thermodynamical equilibrium. It is tacitly used in phenomenological "open systems quantum mechanics". The *reduction* of entropy in a process of symmetry breaking, on the other hand, is usually very small when compared with thermodynamic entropy, but it may be cosmologically essential when, for example, it leads to new Goldstone type particles that usually possess an enormous entropy capacity (see Sect. 6.1 of TD).

Another novel consequence of quantum theory that regards the arrow of time is the entropy bound that governs gravitational contraction. It is characterized by the

Bekenstein–Hawking black hole entropy, given by

$$S_{BH} = 4\pi \frac{k_B G M^2}{\hbar c} \tag{3}$$

for spherical and electrically neutral black holes. Here, G is the gravitational constant. The fact that $S_{\rm BH}$ is quadratic in the mass M indicates that it must describe some kind of correlations. According to classical general relativity, spacetime geometry is regular at the black hole horizon, while there has to be a future singularity inside. However, the interior cannot causally affect the external region any more: it must for all times remain in the future of all external observers. This leaves much freedom for the unknowable physics inside. In particular, *quantum* gravity does not allow one to distinguish between past and future singularities any more (see below). Therefore, one can only postulate a Weyl tensor condition on *all* space-like singularities. Such a time-symmetric condition is not only compatible with all observations – it may even prevent black hole interiors and horizons to form (thus avoiding any genuine information loss paradox) [10, 11].

Most of these genuine quantum aspects of the cosmic arrow of time have so far received little attention – perhaps because they seem to depend on the interpretation of the quantum formalism. Cosmological models are mostly presented in classical terms rather than in terms of quantum states (superpositions). In particular, arguments based on Feynman's path integral often replace this integral, which describes a superposition of paths (precisely equivalent to a wave function [12]) by an *ensemble* of paths in classical configuration space. Selecting a subensemble or an individual path from them is nonetheless equivalent to a time-asymmetric collapse of the wave function. Similar objections apply to tunneling probabilities, since any decay process must quantum mechanically be described as a coherent superposition of different decay times as long as the corresponding partial waves are not irreversibly decohered from one another (thereby letting decay events appear to be "real" rather than virtual – Sect. 5 of TD).

A consistent quantum description requires that classical general relativity is replaced by quantum gravity. This does not necessarily require a complete understanding of this theory. While the problem of the arrow of time can probably be finally answered only in an ultimate theory, the meaning and validity of existing proposals (such as in string theories) have remained highly speculative as yet. Standard quantization of the canonical form [13] of General Relativity in the Schrödinger picture, on the other hand, leads to the Wheeler–DeWitt equation (or Hamiltonian quantum constraint) [14],

$$H\psi = 0, \tag{4}$$

which may be expected to form an effective theory of quantum gravity at "low" (that is, normal) energies. The wave functional Ψ depends on spatial geometries and matter fields on arbitrary simultaneities. Since the Schrödinger equation now takes the form $\partial\Psi/\partial t = 0$, there exists no time parameter any more that could be used to formulate a direction in time. This "timelessness" has occasionally been regarded as a severe blow to this approach, although it must apply to all quantum theories that are reparametrization invariant in their classical form. However, the physical concept of time – and even its arrow – can be recovered and understood in a satisfactory way under very reasonable assumptions [15, 16].

The first important observation for this purpose is that the Wheeler–DeWitt equation for Friedmann type universes is globally of hyperbolic type, with a time-like variable $\alpha := \ln\, a$, where a is the cosmic expansion parameter [17–20]. This fact defines an intrinsic "initial" value problem in α or a, for example at the big bang ($\alpha = -\infty$), which would in configurations space be identical with a big crunch. The Wheeler–DeWitt equation is drastically asymmetric under a change of sign of α, thus suggesting an asymmetric solution without explicitly postulating it by means of asymmetric boundary conditions. The second step for recovering conventional time is a Born–Oppenheimer expansion in terms of the inverse Planck mass, which equals 1.3×10^{19} proton masses [21, 22]. This mass characterizes all geometric degrees of freedom. The expansion leads to an approximately autonomous evolution of partial Wheeler–DeWitt wave functions for the matter degrees of freedom along WKB trajectories that are defined in most regions of the configuration space of geometries. This is analogous to the adiabatic evolution of electron wave functions along classical orbits of the heavy nuclei in large molecules. This evolution has precisely the form of a time-dependent Schrödinger equation (plus very small corrections) [23]. The concept of time recovered in this way represents arbitrary time coordinates for all possible foliations, and independently for all dynamically arising quasi-classical spacetimes (branches).

Note that this WKB approximation does not by itself justify an *ensemble* of trajectories, since it preserves the global superposition that they form. Similarly, small molecules (for which the positions of nuclei are *not* decohered to become quasi-classical variables) are known to exist in energy eigenstates (wave functions) in spite of the validity of the Born-Oppenheimer approximation. However, since observers would also possess different states in the different autonomous partial waves for the universe, they can observe only their own "branch" as an apparently evolving quantum world. The global intrinsic dynamics would be required, though, in order to dynamically *derive* the initial conditions for all partial Schrödinger wave functions that have to be used in the WKB region of geometry (at some distance from the big bang).

According to arguments used in loop quantum cosmology, the Wheeler–DeWitt equation (in this theory replaced by a difference equation with respect to a) can be continued through $a = 0$ to negative values of a [24, 25]. The configuration space of three-geometries is in this way duplicated by letting the volume measure assume negative values (turning space "inside out" while going through $a = 0$). Since the Hamiltonian does not depend on the newly invented sign of a, however, the Wheeler–DeWitt wave function must be expected to be symmetric under this parity transformation, too, in the absence of any artificial boundary condition. Its continuation would then have to be interpreted as an enlarged superposition of components that are all individually experienced as *expanding* universes. Since their

WKB times, which represent classical times, can *not* be continued through $a = 0$, where the WKB approximation breaks down, the interpretation of negative values of a as representing pre-big-bang times is highly questionable. The fundamental arrow, including its consequence of decoherence with respect to a even outside the validity of a WKB approximation, must depend on some low entropy (no entanglement) "initial" condition in this time-like variable for all other ("spacelike") degrees of freedom that occur as physical arguments of the Wheeler–DeWitt wave function. It would be hard to understand how the low entropy state at $a = 0$ could have been "preceded" by an even lower entropy at $a < 0$ in order to avoid a reversal of the thermodynamical arrow in the classical picture of an oscillating universe.

In spite of the success in recovering physical time for the autonomous Everett branches that represent quasi-classical spacetimes, "timelessness" has recently become a hot issue that is based on some severe misunderstandings. It has even been used as a motivation to present obscure and speculative solutions to this non-existing problem. I will, therefore, now give a brief review of *different* concepts of timelessness that have been discussed and confused in this connection.

4 A Brief History of Timelessness

Newton described planetary motions in the form $r(t),\ \varphi(t)$ that required a concept of absolute time. He concluded from his laws, that absolute time t can be read from appropriate clocks, such as the rotation of the Earth, $\alpha = \omega t$. Elimination of t from the first two functions leads to Kepler's orbits $r(\varphi)$. Similarly, its elimination from all three functions leads to a clock dependence $r(\alpha)$ and $\varphi(\alpha)$. This trivial elimination of time has recently been used by some authors to argue that one should "forget time" in all dynamical considerations [26]. However, this argument completely neglects the fact that it is precisely Newton's time that simplifies his laws of motion, as has been clearly emphasized by Henri Poincaré. So, in Newtonian physics there is a preferred time parameter that could indeed be interpreted as representing "absolute" time.

The concept of absolute time was not only questioned for philosophical reasons by Leibniz and Mach, it also lost its empirical justification in general relativity. In special relativity, absolute time is replaced by an absolute spacetime metric that still defines path-dependent proper times. According to the principle of relativity, they control all physical motions in the same preferred way as Newtonian time did in non-relativistic physics. In particular, local clocks *measure* proper times along their world lines, while the spacetime metric is assumed to exist even in the absence of physical clocks.

In general relativity, the spatial metric defined on arbitrary simultaneities becomes itself a dynamical object [13] – just as any matter field. Its evolution gives rise to a succession of spatial curvatures that defines a foliation of spacetime. It can be parametrized by an arbitrarily chosen time coordinate, but there is no *preferred* coordinate or time parameter any more. Barbour has discussed this

Machian property, which he called timelessness, in great detail, including many consequences that were historically important [27]. However, the arising metric still defines proper times for all world lines (Wheeler's *many-fingered time*), and the evolving spatial metric can itself be regarded as a many-fingered *physical* clock [28]. Although there are many different time-like foliations of the same spacetime, each one defines a parametrizable succession of states, and this dynamical construction allows in general the formulation of a unique initial value problem – hence an initial condition of low entropy. There are also mathematically consistent non-relativistic Machian ("relational") theories [29].

The complete absence of any time parameter from the Wheeler–DeWitt wave function (genuine timelessness), discussed in the previous section, is a specific *quantum* property: it is a consequence of the fact that in quantum theory there are no trajectories (in configuration space) that could be parametrized. Hence, in quantum gravity there are no classical spacetimes that could give rise to a time-like foliation. There is only a probability amplitude for spatial geometries (many-fingered physical clocks) entangled with matter fields [30]. For the same reason, the recently proposed concept of "relational observables" [26] is inappropriate, since it is based on a classical concept of orbits, required to define such relations between variables. Entanglement describes also the decoherence of macroscopically different geometries from one another if matter is regarded as an environment to geometry [15, 16]. Among the parameters characterizing these spatial geometries is the "intrinsic time" $\alpha = \ln a$. It is remarkable that this genuine timelessness (the inapplicability of *any* external time parameter) was known before its weaker classical versions were discussed under this ambitious name. Unfortunately, it seems to have initially been mostly regarded as a merely formal problem. The reason may be that early physicists working on quantum gravity did not take the Wheeler–DeWitt wave function seriously as representing reality. They either used semi-classical approximations for its interpretation (even where they were not justified), or they preferred a Heisenberg picture, in which the problem is less obvious [31].

References

1. H.D. Zeh, *The Physical Basis of the Direction of Time*, 5th edn. (Springer, Berlin, 2007) (cited as TD). See also www.time-direction.de
2. H.D. Zeh, Entropy **7**(4), 199 (2005)
3. H.D. Zeh, **8**(2), 44 (2006), and references therein
4. H. Price, *Time's Arrow & Archimedes' Point: A View from Nowhen* (Oxford University Press, 1996)
5. E. Borel, *Le hasard* (Alcan, Paris, 1924)
6. C.H. Bennett, Sci. Am. **257**, 108 (1987)
7. E. Lubkin, J. Theor. Phys. **26**, 523 (1987)
8. D. Lewis, *Philosophical Papers*, vol. II (Oxford University Press, 1986)
9. J.B. Barbour, Classical Quant. Grav. **11**, 2853 (1994)
10. C. Kiefer, H.D. Zeh, Phys. Rev. **D51**, 4145 (1995)

11. H.D. Zeh, Phys. Lett. **A347**, 1 (2005); see also C. Kiefer's contribution to this volume
12. F.J. Dyson, Phys. Rev. **75**, 486 (1949)
13. R. Arnowitt, S. Deser, C.W. Misner, in *Gravitation – An Introduction to Current Research*, ed. by L. Witten. The Dynamics of General Relativity (Wiley, New York, 1962)
14. B.S. DeWitt, Phys. Rev. **160**, 1113 (1967)
15. For a review see: C. Kiefer, *Quantum Gravity*, 2nd edn. (Oxford University Press, 2007)
16. C. Kiefer (2009), arXiv:0909.3767
17. J.J. Halliwell, S.W. Hawking, Phys. Rev. **D31**, 1777 (1985)
18. H.D. Zeh, Phys. Lett. **A116**, 9 (1986)
19. H.D. Zeh, **A126**, 311 (1988)
20. D. Giulini, Phys. Rev. **D51**, 5630 (1995)
21. T. Banks, Nucl. Phys. **B249**, 332 (1985)
22. R. Brout, G. Venturi, Phys. Rev. **D39**, 2436 (1989)
23. C. Kiefer, Phys. Rev. **D46**, 1658 (1992)
24. M. Bojowald, Gen. Rel. Grav. **35**, 1877 (2003)
25. M. Bojowald, Sci. Am., **299**, 44 (2008)
26. C. Rovelli (2009), arXiv:0903.3832
27. J.B. Barbour, *The End of Time* (Weidenfels & Nicolson, London, 1999)
28. R.F. Baierlein, D.H. Sharp, J.A. Wheeler, Phys. Rev. **126**, 1864 (1962)
29. J.B. Barbour, B. Bertotti, Proc. R. Soc. London **A382**, 295 (1982)
30. D.N. Page, W.K. Wootters, Phys. Rev. **D27**, 2885 (1983)
31. See, for example, C.J. Isham, in *Integrable systems, quantum groups, and quantum field theory*, ed. by L.A. Ibort, M.A. Rodriguez. (Kluwer, Dordrecht, 1993)

Index

L. Mersini-Houghton and R. Vaas (eds.), *The Arrows of Time*, Fundamental Theories of Physics 172, DOI 10.1007/978-3-642-23259-6,

Kolibri 360 GmbH
Ferdinand-Jühlke-Straße 7
99095 Erfurt, Deutschland
produktsicherheit@kolibri360.de